THIRD EDITION

Student Solutions Manual for Kleinbaum, Kupper, Muller, and Nizam's
Applied Regression Analysis and Other Multivariable Methods

Aaron T. Curns
American SIDS Institute

Azhar Nizam
Emory University

Duxbury Press
An Imprint of Brooks/Cole Publishing Company
I(T)P® An International Thomson Publishing Company

Pacific Grove • Albany • Belmont • Bonn • Boston • Cincinnati • Detroit • Johannesburg • London
Madrid • Melbourne • Mexico City • New York • Paris • Singapore • Tokyo • Toronto • Washington

Sponsoring Editor: *Cynthia Mazow*
Editorial Assistant: *Rita Jaramillo*
Production: *Dorothy Bell*
Printing and Binding: *Edwards Brothers, Ann Arbor*

COPYRIGHT © 1998 by Brooks/Cole Publishing Company
A division of International Thomson Publishing Inc.
I(T)P The ITP logo is a registered trademark under license.

Duxbury Press and the leaf logo are registered in the U.S. Patent and Trademark Office.

For more information, contact Duxbury Press at 10 Davis Drive, Belmont, CA 94002, or electronically at http://www.thomson.com/duxbury.html

BROOKS/COLE PUBLISHING COMPANY
511 Forest Lodge Road
Pacific Grove, CA 93950
USA

International Thomson Editores
Seneca 53
Col. Polanco
11560 México, D. F., México

International Thomson Publishing Europe
Berkshire House 168-173
High Holborn
London WC1V 7AA
England

International Thomson Publishing GmbH
Königswinterer Strasse 418
53227 Bonn
Germany

Thomas Nelson Australia
102 Dodds Street
South Melbourne, 3205
Victoria, Australia

International Thomson Publishing Asia
221 Henderson Road
#05-10 Henderson Building
Singapore 0315

Nelson Canada
1120 Birchmount Road
Scarborough, Ontario
Canada M1K 5G4

International Thomson Publishing Japan
Hirakawacho Kyowa Building, 3F
2-2-1 Hirakawacho
Chiyoda-ku, Tokyo 102
Japan

All rights reserved. No part of this work may be reproduced, stored in a retrieval system, or transcribed, in any form or by any means—electronic, mechanical, photocopying, recording, or otherwise—without the prior written permission of the publisher, Brooks/Cole Publishing Company, Pacific Grove, California 93950.

Printed in the United States of America
10 9 8 7 6 5 4 3 2 1
ISBN 0-534-20913-0

Contents

Chapter 5	1
Chapter 6	10
Chapter 7	16
Chapter 8	18
Chapter 9	20
Chapter 10	23
Chapter 11	27
Chapter 12	30
Chapter 13	33
Chapter 14	40
Chapter 15	46
Chapter 16	48
Chapter 17	52
Chapter 18	58
Chapter 19	60
Chapter 20	64
Chapter 21	68
Chapter 22	70
Chapter 23	71
Chapter 24	72

Preface

Solutions are provided in this manual for the odd-numbered problems in the third edition of the textbook. The solutions are generally quite detailed. However, the reader should note that detailed computations are not presented whenever computer output from the textbook can be used directly to obtain a result. We would like to acknowledge the helpful contributions of Seoung Lee and Ryan Holman.

Dedication
For Carrie and Alastor
For Janet, Zainab, and Sohail
-for their love, patience, and support.

Chapter 5

1. a Dry weight (Y) does increase with increasing Age (X) but the scatter diagram illustrates that the relationship may not be linear. An exponential relationship between X and Y may better fit the data.

Log dry weight (Z) increases linearly with increasing Age (X). The scatter diagram illustrates an almost perfect linear relationship between the independent variable X and the dependent variable Z.

Ch.5 Q.1.a   **Ch.5 Q.1.b.**

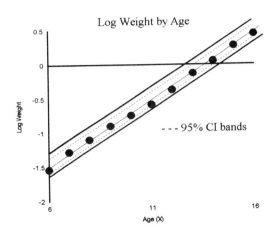

b $Y = \beta_0 + \beta_1 X + E$ $Z = \beta'_0 + \beta'_1 X + E$

c Using the SAS output provided for Problem 1:

$\hat{\beta}_0 = -1.885$ $\hat{\beta}_1 = 0.235$ $\hat{\beta}'_0 = -2.689$ $\hat{\beta}'_1 = 0.196$.

The least squares estimates for each regression line are:

$\hat{Y} = -1.885 + 0.235X$ $\hat{Z} = -2.689 + 0.196X$.

d The regression line for Log_{10} dry weight regressed on Age has a better fit. It is more appropriate to run a linear regression of Z on X because there is an obvious linear relationship between Log_{10} dry weight and Age.

    e To make the interval computation easier, we can use the SAS output provided for our calculations. From the SAS output for Log_{10} dry weight regressed on Age:

$$\hat{\beta}_1' = 0.196 \quad S_{\hat{\beta}_1'} = 0.003. \quad \hat{\beta}_0' = -2.689 \quad S_{\hat{\beta}_0'} = 0.031.$$

95% Confidence Interval for $\hat{\beta}_1' \; : \; \hat{\beta}_1' \pm t_{n-2, 0.975} * S_{\hat{\beta}_1'}$

$0.196 \pm 2.262 * 0.003 = (0.189, 0.203)$

We are 95% confident that the true slope is between 0.189 and 0.203. Since the interval does not contain zero we reject the null hypothesis that the slope equals zero at $\alpha = 0.05$.

95% Confidence Interval for $\hat{\beta}_0' \; : \; \hat{\beta}_0' \pm t_{n-2, 0.975} * S_{\hat{\beta}_0'}$

$-2.689 \pm 2.262 * 0.031 = (-2.759, -2.619)$

We are 95% confident that the true intercept is between -2.759 and -2.619. Since the interval does not contain zero we reject the null hypothesis that the slope equals zero at $\alpha = 0.05$.

    f From the SAS output we see that observation 3 contains information about an eight-day-old chick. Confidence intervals are included in the output. We are 95% confident that the true mean Log_{10} dry weight of an eight-day-old chick lies between -1.149 and -1.096.

3. **a** See graph. The relationship between time (Y) and Inc (X) does not appear to be linear.

    b From the SAS output: $\hat{\beta}_0 = 19.626 \quad \hat{\beta}_1 = 0.0007$

    c $\hat{Y} = 19.626 + 0.0007X$. The regression line fits the data poorly.

    d The linearity assumption is not met.

    e $H_o: \beta_1 = 0 \quad\quad H_a: \beta_1 \neq 0 \quad\quad$ TS: T = 2.023, p = 0.0582 (from SAS output)

We do not reject H_o since the p-value > 0.05. There is not evidence to suggest that the slope is different from zero; there is no significant straight-line relationship between time(Y) and Inc (X).

    f The scatter plot suggests that a parabola would better fit the data.

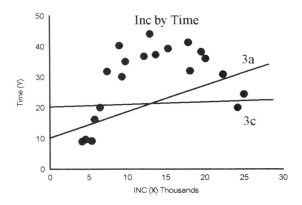

5. **a** $\hat{Y} = 2.174 + 1.177X$. See graph for plot. The line fits the data well.

   b No.

   c $H_o: \beta_1 = 0 \qquad H_a: \beta_1 \neq 0 \qquad$ TS: T = 13.5 p = 0.0001 (from SAS output)

   Since the p-value < 0.05 we reject H_o and conclude that the slope is not equal to 0. There is a significant linear relationship between TVEXP and VOTE.

   d Using the estimated regression line $\hat{Y} = 2.174 + 1.177X$: $\hat{Y}$ equals 45.711 when X = 36.99.

   $$S_{\hat{Y}_{X_0}} = S_{YX}\sqrt{\frac{1}{n} + \frac{(X_0 - \overline{X})^2}{S_X^2(n-1)}} = 3.332 * \sqrt{1/20 + 0} = 0.745$$

   $H_o: \mu_{YX_0} = 45 \qquad H_a: \mu_{YX_0} \neq 45 \qquad$ TS: $T_{18} = \dfrac{\hat{Y}_{X_0} - \mu_{Y^0|X_0}}{S_{\hat{Y}_{X_0}}} = \dfrac{45.711 - 45}{0.745} = 0.954$, with

   $0.3 < p < 0.5$.

   Since the p-value > 0.05 we do not reject H_o that $\mu_{YX_0} = 45$. The true mean voter turnout is not significantly different from 45%.

   e $\hat{Y}_{X_0} \pm t_{n-2} S_{\hat{Y}_{X_0}} = 45.711 \pm 2.101 * 0.745 = (44.146, 47.276)$. We are 95% confident that the mean voter turnout is between 44.146 and 47.276 when TV expenditures are 37% of total campaign expenditures.

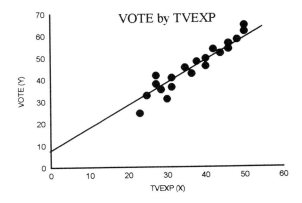

7. **a** For Y_1 regressed on X: from the SAS output: $\hat{\beta}_0 = -122.345$ $\hat{\beta}_1 = 6.227$

$\hat{Y} = -122.345 + 6.227X$. See graph for estimated line.

For Y_2 regressed on X. From the SAS output: $\hat{\beta}_0 = -1.697$ $\hat{\beta}_1 = 0.299$

$\hat{Y} = -1.697 + 0.299X$. See graph for estimated line.

b Y_2 regressed on X.

c $H_0: \beta_1 = 1$ $H_a: \beta_1 \neq 1$ TS: $T = \dfrac{\hat{\beta}_1 - \beta_1^{(0)}}{\dfrac{s_{Y|X}}{s_X\sqrt{n-1}}} = \dfrac{0.299 - 1}{\dfrac{0.811}{15.765\sqrt{18}}} = \dfrac{-0.701}{0.0121} = -57.934$

Critical value: $t_{17} \sim 2.898$ under H_a at $\alpha = 0.01$
We see that $|T| = 57.934$ which exceeds the critical value so we reject H_0 at $\alpha = 0.01$ conclude the slope is not equal to one.

d For the 99% CI: $\hat{\beta}_1 \pm t_{n-2} * SE_{\hat{\beta}_1} = 0.299 \pm 2.898 * 0.012 = 0.299 \pm 0.035 = (0.264, 0.334)$.

We can be 99% confident that the true slope is between 0.264 and 0.334.

e From the SAS output we see that observation 4 has X = 45. The estimated mean value for $Y_2 = 11.748$ with a 95% confidence interval between 11.333 and 12.163.

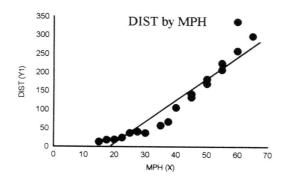

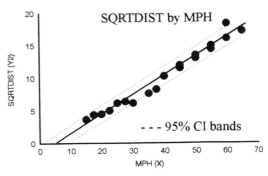

9. **a** $\hat{\beta}_0 = 2.936$ $\hat{\beta}_1 = -1.785$

b $\hat{Y} = 2.936 - 1.785X$. The line fits the data well.

c See chart.

d $\hat{Y} = 2.936 - 1.785X$

where $\hat{Y}' = \log_{10}\hat{Y}$ and $\hat{X}' = \log_{10}\hat{X}$

so: $\log_{10}\hat{Y} = 2.936 - 1.785\log_{10}\hat{X}$

$\log_{10}\hat{Y} + \log_{10}\hat{X}^{1.785} = 2.936$

$\log_{10}(\hat{Y} * \hat{X}^{1.785}) = 2.936$

$\hat{Y} * \hat{X}^{1.785} = 10^{2.936}$

$\hat{Y} = 862.979\hat{X}^{-1.785}$

e   From the SAS output we can use the predicted values of $\hat{Y}$ as well as $S_{\hat{Y}}$ for the Max and Min doses.

99% CI for $\hat{Y}$ given X = 14.125 equals: $\hat{Y} \pm t_{10, 0.995} S_{\hat{Y}_{X_{14.125}}} = 0.883 \pm 3.169(0.027) =$

(0.797, 0.969) then transform $(10^{0.797}, 10^{0.969}) = (6.266, 9.311)$

99% CI for $\hat{Y}$ given X = 1.514 equals: $2.615 \pm 3.169(0.034)$

$= (10^{2.507}, 10^{2.723}) = (321.366, 528.445)$

f   One could plot the transformed data (X′,Y′) and then draw the estimated regression line on the plot. Then one could compare the fit of the estimated line of (X,Y) versus that of (X′,Y′). If one did this comparison, the straight line regression of (X,Y) gives a better fit.

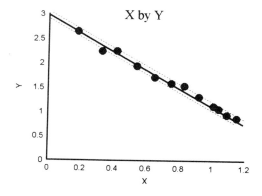

11. a $\hat{\beta}_0 = 3.707$ $\hat{\beta}_1 = -0.012$ $\hat{Y} = 3.707 - 0.012X$. See chart for estimated line.

b $H_o: \beta_1 = 0$ $H_a: \beta_1 \neq 0$ TS: T = -8.684   p = 0.001 (from SAS output)

Since the p-value < 0.05 we reject H_o and conclude that the slope is statistically significant.

c   Including the data from the three experiments rather than just using the average values would provide more information and might improve the sensitivity of the analysis.

d   For X = 100, 90% CI equals: $\bar{Y} + \hat{B}_1(X - \bar{X}) \pm t_{4,\,0.95} * S_{Y|X} \sqrt{\frac{1}{n} + \frac{(X-\bar{X})^2}{n(S_X^2)}}$

$= 3.075 - 0.012(100 - 51.233) \pm 2.132 * 0.151 \sqrt{\frac{1}{6} + \frac{(100-51.233)^2}{5(2264.859)}}$

$= 2.490 \pm 2.132 * 0.585 = (1.243, 3.737)$

e   It is inappropriate since the estimated model is relying on data that does not include any information for average growth rate when exposed to a gas with a molecular weight of 200.

f   The choice of X values are not uniformly distributed in the experiment. There are large gaps between the X values of 39.9, 83.8, and 131.3. This may result in a fitted line subject to inaccuracies for predicting Y based on X.

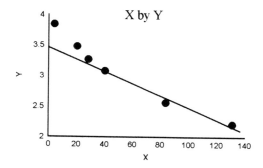

13. a   See chart.

b   From the SAS output: $\hat{\beta}_0 = 0.116$, $\hat{\beta}_1 = 0.005$

c $H_o: \beta_1 = 0$ $H_r: \beta_1 \neq 0$ TS: T = 0.811   p = 0.433 (from SAS output)

Since the p-value > 0.10 we accept H_o and conclude there is not sufficent evidence suggesting the slope is not equal to zero.

d $H_o: \beta_0 = 0$ $H_a: \beta_0 \neq 0$ TS: T = 0.046   p = 0.964 (from SAS output)

Since the p-value > 0.10 we accept H_o and conclude there is not sufficent evidence suggesting the slope is not equal to zero.

e $\hat{Y} = 0.116 + 0.005X$. See chart. The line is nearly horizontal implying there is no linear relationship between body weight and latency to seizure.

f The line does not differ from the line plotted in (e). The evidence suggests that there is no significant linear relationship. Determining a well fitting line is difficult given the dispersion of the data.

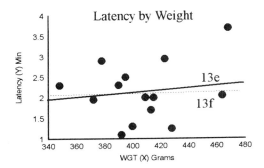

15. a See chart. As X increases the dispersion of Y increases.

b $\hat{\beta}_0 = -2.546$ $\hat{\beta}_1 = 0.032$ $\hat{Y} = -2.546 + 0.032X$

c See chart. This chart implies a better linear relationship between X and Y.

d $\hat{\beta}_0 = -6.532$ $\hat{\beta}_1 = 1.430$ $\hat{Y} = -6.532 + 1.430X$

e The natural log transformation provides the best representation. The natural log plot illustrates the linear relationship better and the dispersion of the data is more similar at each level of toluene exposure. The first plot indicates that there may be a violation of homoscedasticity for the untransformed data.

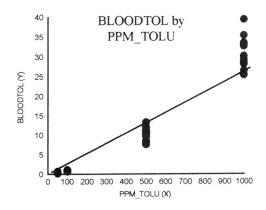

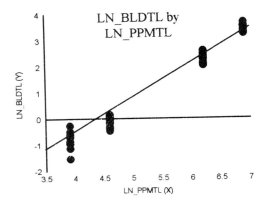

17. a Yes it does.

b $\hat{\beta}_0 = 1.643$ $\hat{\beta}_1 = 1.057$ $\hat{Y} = 1.643 + 1.057X$. See chart. Yes the line appears to fit the data well.

c 95 % CI for β_1 (using the SAS output): $1.057 \pm 2.776 * 0.172 = (0.580, 1.534)$. Yes, because the 95% CI does not include the null value of zero indicating that there is a significant linear relationship at $\alpha = 0.05$.

d No it is not be appropriate since the data used to estimate the regression line does not inlude a $10 million advertising expenditure in its range.

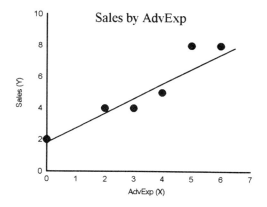

19. a There may be a slight negative linear relationship.

b $Y = \beta_0 + \beta_1 X + E$. $\hat{\beta}_0 = 76.008$ $\hat{\beta}_1 = -0.015$.

The baseline OWNEROCC = 76% and as OWNCOST increases by $1,000 the percentage of OWNEROCC decreases by ~ 2%.

c $\hat{Y} = 76.008 - 0.015X$. See chart for line. The line fits the data well.

d $H_o: \beta_1 = 0$ $H_a: \beta_1 \neq 0$ TS: T = -2.607   p = 0.0155 (from SAS output)

Since the p-value < 0.05 we reject H_o and conclude that the slope is not equal to zero. There is a significant linear relationship between OWNEROCC and OWNCOST.

e   Using the SAS output, the 95% CI for β_1 equals: $-0.015 \pm 2.064 * 0.006 = (-0.027, -0.003)$.

We are 95% confident that the true slope is between -0.027 and -0.003. Again we conclude that the slope is not equal to zero since the interval does not contain zero.

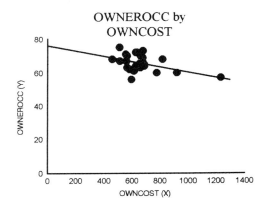

Chapter 6

1. a **(1)** To calculate r we need S_x, S_y, and $\hat{\beta}_1$: $\qquad S_x = 3.317 \qquad S_y = 0.904$

$$\hat{\beta}_1 = 0.235$$

$$r = \frac{S_X}{S_Y}\hat{\beta}_1 = \frac{3.317}{0.904} * 0.235 = 0.862 \ .$$

Alternatively we can take the square root of r^2 provided in the SAS output; the sign of r is the same as the sign of $\hat{\beta}_1$:

$$r = \sqrt{r^2} = \sqrt{0.7442} = 0.863 \ .$$

(2) Using the SAS output provided: $r = \sqrt{r^2} = \sqrt{0.9983} = 0.999$ (the sign of r is the same as the sign of $\hat{\beta}_1$)

b **(1)** $\quad \frac{1}{2} \ln\frac{1 + r}{1 - r} \pm \frac{Z_{1-\alpha/2}}{\sqrt{n - 3}} = \frac{1}{2} \ln\frac{1 + 0.863}{1 - 0.863} \pm \frac{1.96}{\sqrt{8}} \quad 1.305 \pm 0.693 = (0.612, 1.998).$

This interval needs to be transformed back for ρ values. One could transform by hand or use Table A-5 to get approximate values. Hand transformation values = (0.546, 0.964). Table values = (0.546, 0.964).

(2) $\quad \frac{1}{2} \ln\frac{1 + 0.999}{1 - 0.999} \pm \frac{1.96}{\sqrt{8}} = 3.800 \pm 0.693 = (3.107, 4.493).$

Using Table A-5 the 95% CI equals (0.996, 1.000).

c **i.** Hand calculation: $r^2 = 0.863^2 = 0.745 \qquad$ SAS output: $r^2 = 0.744$. 74% percent of the variation in Y is explained with the help of X.
ii. Hand calculation: $r^2 = 0.999^2 = 0.998 \qquad$ SAS output: $r^2 = 0.998$. 99% percent of the variation in Y is explained with the help of X.

d The regression using Log_{10} Dry Weight appears to fit the data better. The model explains a much greater proportion of the total variation in the dependent variable than does the regression of dry weight on age. This conclusion agrees with the conclusion for Problem 1(d) of Chapter 5.

3. a $S_x = 5.750 \qquad S_y = 5.750$. To calculate $\hat{\beta}_1$ use formula shown in 2: $\qquad \hat{\beta}_1 = 0.980$

$$r = \frac{S_X}{S_Y}\hat{\beta}_1 = \frac{5.750}{5.750} * 0.980 = 0.980$$

b Substituting $r(S_y/S_x)$ and $\frac{(n-1)}{(n-2)}(S_Y^2 - \hat{\beta}_1^2 S_X^2)$ for $\hat{\beta}_1$ and $S_{Y|X}^2$ in the

formula $T = \dfrac{\hat{\beta}_1}{\dfrac{S_{Y|X}}{S_X\sqrt{n-1}}}$, we

get $\dfrac{r\dfrac{S_Y}{S_X}S_X\sqrt{n-1}}{\sqrt{\dfrac{n-1}{n-2}\left[S_Y^2 - r^2\dfrac{S_Y^2}{S_X^2}S_X^2\right]}} = \dfrac{rS_Y\sqrt{n-1}}{\sqrt{\dfrac{n-1}{n-2}}\sqrt{S_Y^2 - r^2 S_Y^2}} = \dfrac{rS_Y\sqrt{n-1}\sqrt{n-2}}{\sqrt{n-1}\sqrt{S_Y^2(1-r^2)}} = T'$

c $T' = \dfrac{r\sqrt{n-2}}{\sqrt{1-r^2}} = \dfrac{0.980\sqrt{8}}{\sqrt{1-0.980^2}} = 13.929$ which has the t distribution with 8 df under H_o: $\rho = 0$. The resulting p-value for this test is less than 0.001 therefore we reject H_o.

d The graph of Y vs X does not illustrate a linear relationship.

5. a From the SAS output for SBP on AGE: $r^2 = 0.601$ $r = 0.775$ (the sign of r is the same as the sign of $\hat{\beta}_1$). 60% of the variation in SBP (Y) is explained by AGE (X).

b $\dfrac{1}{2}\ln\dfrac{1+0.775}{1-0.775} \pm \dfrac{2.576}{\sqrt{32-3}} = 1.033 \pm 0.478 = (0.554, 1.511)$ Using Table A-5 to transform this interval, we obtain the following 99% CI for ρ: (0.504, 0.907). Since $\rho = 0$ is not included in the interval we reject H_o: $\rho = 0$ at $\alpha = 0.01$.

7. a From the SAS output for TIME on INC: $r^2 = 0.1853$ $r = 0.430$ (sign of r is the same as the sign of $\hat{\beta}_1$). 19% of the variation in SBP (Y) is explained by AGE (X).

b H_0: $\rho = 0$ H_a: $\rho \neq 0$

$T = \dfrac{0.430\sqrt{20-2}}{\sqrt{1-0.1853}} = \dfrac{1.824}{0.903} = 2.02$

Critical value: $t_{18, 0.975} = 2.101$

At $\alpha = 0.05$, since $|T| <$ critical value, we would not reject H_o and conclude that there is not evidence suggesting that $\rho \neq 0$.

c $\dfrac{1}{2}\ln\dfrac{1+0.430}{1-0.430} \pm \dfrac{1.96}{\sqrt{20-3}} = 0.460 \pm 0.475 = (-0.015, 0.935)$ Using Table A-5 to

transform this interval, we obtain the following 95% CI for ρ: $(-0.015, 0.733)$. Since $\rho = 0$ is included in the interval we do not reject H_o: $\rho = 0$ at $\alpha = 0.05$ which agrees with the conclusion in (b).

9. a From the SAS output for VOTE (Y) on TVEXP (X): $r^2 = 0.9101$ $r = 0.954$ (the sign of r is the same as the sign of $\hat{\beta}_1$). 91% of the variation in Y is explained by X.

b $H_0: \rho = 0$ $H_a: \rho \neq 0$ Critical value: $t_{18,\,0.975} = 2.101$

$$T = \dfrac{0.954\sqrt{20-2}}{\sqrt{1-0.9101}} = \dfrac{4.047}{0.300} = 13.49$$

At $\alpha = 0.05$ we would reject H_o and conclude that there is evidence suggesting that $\rho \neq 0$.

c $\dfrac{1}{2}\ln\dfrac{1+0.954}{1-0.954} \pm \dfrac{1.96}{\sqrt{20-3}} = 1.874 \pm 0.475 = (1.399, 2.349)$ Using Table A-5 to

transform this interval, we obtain a 95% CI for ρ: $(0.885, 0.982)$. Since $\rho = 0$ is not included in the interval we reject H_o: $\rho = 0$ at $\alpha = 0.05$ which agrees with the conclusion in (b).

11. a From the SAS output for SQRT DIST (Y_2) on MPH (X): $r^2 = 0.9728$ $r = 0.986$ (the sign of r is the same as the sign of $\hat{\beta}_1$). 98% of the variation in Y_2 is explained by X.

b $H_0: \rho = 0$ $H_a: \rho \neq 0$

$$T = \dfrac{0.986\sqrt{19-2}}{\sqrt{1-0.9728}} = \dfrac{4.065}{0.165} = 24.640$$

Critical value: $t_{17,\,0.975} = 2.110$

At $\alpha = 0.05$, since $|T| >$ critical value, we would reject H_o and conclude that there is evidence suggesting that $\rho \neq 0$.

  c $\dfrac{1}{2} \ln \dfrac{1 + 0.986}{1 - 0.986} \pm \dfrac{1.96}{\sqrt{19 - 3}} = 2.477 \pm 0.490 = (1.987, 2.967)$ Using Table A-5 to

    transform this interval we obtain the following 95% CI for ρ: (0.963, 0.995). Since $\rho = 0$ is not included in the interval we reject H_o: $\rho = 0$ at $\alpha = 0.05$ which agrees with the conclusion in (b).

13. a Calculation of the sample correlation coefficient between U and V:

$$S_U = 0.411 \qquad S_V = 0.099 \qquad \hat{\beta}_1 = 0.234$$

$$r = \dfrac{S_U}{S_V} \hat{\beta}_1 = \dfrac{0.411}{0.099}\, 0.234 = 0.971$$

  b $\dfrac{1}{2} \ln \dfrac{1 + 0.971}{1 - 0.971} \pm \dfrac{2.576}{\sqrt{10 - 3}} = 2.110 \pm 0.974 = (1.136, 3.084)$ Using Table A-5 to

    transform this interval, we obtain a 99% CI for ρ: (0.813, 0.996). Since $\rho = 0$ is not included in the interval we reject H_o: $\rho = 0$ at $\alpha = 0.01$ and accept H_a : $\rho \neq 0$.

15. a Under the usual straight line regression assumptions, an appropriate large sample test statistic for the one sample test of $H_0: \rho_{1g} = \rho_{2g}$ is:

$$Z = \dfrac{(r_{1g} - r_{2g})\sqrt{n}}{\sqrt{(1 - r_{1g}^2)^2 + (1 - r_{2g}^2)^2 - 2r_{12}^2 - 2[(r_{12} - r_{1g}r_{2g})(1 - r_{1g}^2 - r_{2g}^2 - r_{12}^2)]}} =$$

$$\dfrac{(0.55 - 0.42)\sqrt{121}}{\sqrt{[1-(0.55)^2]^2 + [1-(0.42)^2]^2 - 2(0.70)^2 - 2[(0.70 - (0.55*0.42))(1 - (0.55)^2 - (0.42)^2 - (0.70)^2)]}}$$

= 3.62

The critical region is $|Z| > 1.96$ for a two-tailed test at $\alpha = 0.05$. Since $|Z| = 3.62$ is greater than 1.96, we reject the null hypothesis.

17. a From the SAS output for SAL (Y)* on CGPA (X): $r^2 = 0.9558$ $r = 0.978$. 96% of the variation in Y is explained by X. *includes 16th obs.

  b $H_0: \rho = 0$ $H_a: \rho \neq 0$ Critical value: $t_{14, 0.975} = 2.145$

$$T = \dfrac{0.978\sqrt{16 - 2}}{\sqrt{1 - 0.9558}} = \dfrac{3.659}{0.210} = 17.424$$

At $\alpha = 0.05$ we would reject H_o and conclude that there is evidence suggesting that $\rho \neq 0$.

c $\quad \dfrac{1}{2}\ln\dfrac{1+0.978}{1-0.978} \pm \dfrac{1.96}{\sqrt{16-3}} = 2.249 \pm 0.544 = (1.705, 2.793)$ Using Table A-5 to

transform this interval, we obtain the following 95% CI for ρ: $(0.936, 0.993)$. Since $\rho = 0$ is not included in the interval we reject H_o: $\rho = 0$ at $\alpha = 0.05$ which agrees with the conclusion in (b).

19. a From the SAS output for WAV (Y) on YEAR (X): $\quad r^2 = 0.0310 \quad\quad r = -0.176$. 3% of the variation in Y is explained by X.

b H_0: $\rho = 0 \quad H_a$: $\rho \neq 0$ Critical value: $t_{18, 0.975} = 2.101$

$$T = \dfrac{-0.176\sqrt{20-2}}{\sqrt{1-0.031}} = \dfrac{-0.747}{0.984} = -0.759 \quad |T| = 0.759$$

At $\alpha = 0.05$ we would not reject H_o and conclude that there is not evidence suggesting that $\rho \neq 0$.

c $\quad \dfrac{1}{2}\ln\dfrac{1-0.176}{1+0.176} \pm \dfrac{1.96}{\sqrt{20-3}} = -0.178 \pm 0.475 = (-0.653, 0.297)$ Using Table A-5 to

transform this interval, we obtain the following 95% CI for ρ: $(-0.574, 0.289)$. Since $\rho = 0$ is included in the interval we do not reject H_o: $\rho = 0$ at $\alpha = 0.05$ which agrees with the conclusion in (b).

21. a From the SAS output for LNBLOOD (Y) on LNPPMTL (X): $\quad r^2 = 0.9813 \quad\quad r = 0.991$. 98% of the variation in Y is explained by X.

b H_0: $\rho = 0 \quad H_a$: $\rho \neq 0$ Critical value: $t_{58, 0.975} = 2.0$

$$T = \dfrac{0.991\sqrt{60-2}}{\sqrt{1-0.9813}} = \dfrac{7.547}{0.137} = 55.088$$

At $\alpha = 0.05$ we would reject H_o and conclude that there is evidence suggesting that $\rho \neq 0$.

c $\quad \dfrac{1}{2}\ln\dfrac{1+0.991}{1-0.991} \pm \dfrac{1.96}{\sqrt{60-3}} = 2.700 \pm 0.260 = (2.440, 2.960)$ Using Table A-5 to

transform this interval, we obtain the following 95% CI for ρ: $(0.985, 0.995)$. Since $\rho = 0$ is not included in the interval we reject H_o: $\rho = 0$ at $\alpha = 0.05$ which agrees with the conclusion in (b).

23. a From the SAS output for SALES (Y) on ADVEXP (X): $\quad r^2 = 0.9044 \quad\quad r = 0.951$. 90% of the variation in Y is explained by X.

b H_0: $\rho = 0 \quad H_a$: $\rho \neq 0$ Critical value: $t_{4, 0.975} = 2.776$

$$T = \dfrac{0.951\sqrt{6-2}}{\sqrt{1-0.9044}} = \dfrac{1.902}{0.309} = 6.155$$

At $\alpha = 0.05$ we would reject H_o and conclude that there is evidence suggesting that $\rho \neq 0$.

c $\dfrac{1}{2}\ln\dfrac{1+0.951}{1-0.951} \pm \dfrac{1.96}{\sqrt{6-3}} = 1.842 \pm 1.132 = (0.711, 2.974)$ Using Table A-5 to

transform this interval, we obtain the following 95% CI for ρ: $(0.611, 0.995)$. Since $\rho = 0$ is not included in the interval we reject H_o: $\rho = 0$ at $\alpha = 0.05$ which agrees with the conclusion in (b).

25. a   From the SAS output for OWNOCC (Y) on OWNCOST (X): $r^2 = 0.2207$ $r = 0.470$.
22% of the variation in Y is explained by X.

b H_0: $\rho = 0$ H_a: $\rho \neq 0$ Critical value: $t_{24, 0.975} = 2.064$

$$T = \dfrac{0.470\sqrt{26-2}}{\sqrt{1-0.2207}} = \dfrac{2.303}{0.883} = 2.608$$

At $\alpha = 0.05$ we would reject H_o and conclude that there is evidence suggesting that $\rho \neq 0$.

c $\dfrac{1}{2}\ln\dfrac{1+0.470}{1-0.470} \pm \dfrac{1.96}{\sqrt{26-3}} = 0.51 \pm 0.409 = (0.101, 0.919)$ Using Table A-5 to

transform this interval, we obtain the following 95% CI for ρ: $(0.101, 0.725)$. Since $\rho = 0$ is not included in the interval we reject H_o: $\rho = 0$ at $\alpha = 0.05$ which agrees with the conclusion in (b).

Chapter 7

1. **a** **i**

	df	Sum of Squares	Mean Sum of Squares	F
Regression	1	6.080	6.080	26.207
Residual	9	2.088	0.232	
	10	8.168		

   ii

	df	Sum of Squares	Mean Sum of Squares	F
Regression	1	4.221	4.221	5276.25
Residual	9	0.007	0.0008	
	10	4.228		

   b **i** $H_0: \beta_1 = 0$ $H_a: \beta_1 \neq 0$ Critical Value: $F_{1, 9, 0.95} = 5.12$

   Since 26.607 > 5.12, the critical value at $\alpha = 0.05$, we would reject H_0 and conclude that there is a significant linear relationship of Y on X.

   ii $H_0: \beta_1 = 0$ $H_a: \beta_1 \neq 0$ Critical Value: $F_{1, 9, 0.95} = 5.12$

   Since 5276.25 > 5.12 we would reject H_0 and conclude that there is a significant linear relationship of Z on X at $\alpha = 0.05$.

5. **a**

	df	Sum of Squares	Mean Sum of Squares	F
Regression	1	450.865	450.865	4.093
Residual	18	1982.916	110.162	
	19	2433.781		

   b $H_0: \beta_1 = 0$ $H_a: \beta_1 \neq 0$ Critical Value: $F_{1, 18, 0.95} = 4.41$

   Since 4.093 < 4.41 we would not reject H_0 and conclude that there is not a significant linear relationship of Y on X at $\alpha = 0.05$.

   c $T^2 = (2.023)^2 = 4.093$. The values are virtually the same.

   d The hypotheses for each test are equivalent. As mentioned in the text, the F statistic and T statistic are equivalent after squaring T. Using the information that the tests of hypotheses are equivalent, as are the test statistics, one may infer that the resultant p-values for each test would also be equivalent.

9. **a**

	df	Sum of Squares	Mean Sum of Squares	F
Regression	1	76858486.06	76858486.06	60.758
Residual	28	35419546.54	1264983.805	
	29	112278032.670		

   b $H_0: \beta_1 = 0$ $H_a: \beta_1 \neq 0$ Critical Value: $F_{1, 28, 0.95} = 4.20$

   Since 60.758 > 4.20 we would reject H_0 and conclude that there is a significant linear relationship of Y on X at $\alpha = 0.05$.

13. a

	df	Sum of Squares	Mean Sum of Squares	F
Regression	1	133806354.1	133806354.1	3155.890
Residual	14	593584.922	42398.923	
	15	134399939.0		

b $H_0: \beta_1 = 0$ $H_a: \beta_1 \neq 0$ Critical Value: $F_{1, 14, 0.95} = 4.60$

Since 3155.89 > 4.60 we would reject H_0 and conclude that there is a significant linear relationship of Y on X at $\alpha = 0.05$.

17. a

	df	Sum of Squares	Mean Sum of Squares	F
Regression	1	177.547	177.547	3061.155
Residual	58	3.364	0.058	
	59	180.911		

b $H_0: \beta_1 = 0$ $H_a: \beta_1 \neq 0$ Critical Value: $F_{1, 58, 0.95} = 4.00$

Since 3061.155 > 4.00 we would reject H_0 and conclude that there is a significant linear relationship of Y on X at $\alpha = 0.05$.

21. a

	df	Sum of Squares	Mean Sum of Squares	F
Regression	1	132.626	132.626	6.796
Residual	24	468.336	19.514	
	25	600.962		

b $H_0: \beta_1 = 0$ $H_a: \beta_1 \neq 0$ Critical Value: $F_{1, 24, 0.95} = 4.26$

Since 6.796 > 4.26 we would reject H_0 and conclude that there is a significant linear relationship of Y on X at $\alpha = 0.05$.

Chapter 8

1. a i $\hat{Y} = 45.103 + 1.213(50) + 9.946(1) + 8.592(3.5) = 145.771$

ii $\hat{Y} = 45.103 + 1.213(50) + 9.946(0) + 8.592(3.5) = 135.825$

iii $\hat{Y} = 45.103 + 1.213(50) + 9.946(1) + 8.592(3.0) = 141.475$

As QUET increases from 3.0 to 3.5, average SBP increases by an estimated 4.296 points, from 141.475 to 145.771.

b $R^2 = \dfrac{SSY - SSE}{SSY}$

For SBP on AGE: $R^2 = \dfrac{6425.969 - 2564.338}{6425.969} = 0.601$

For SBP on AGE and SMK: $R^2 = \dfrac{6425.969 - 1736.285}{6425.969} = 0.730$

For SBP on AGE, SMK, and QUET: $R^2 = \dfrac{6425.969 - 1536.143}{6425.969} = 0.761$

The model using AGE and SMK to predict SBP appears to be the best choice. The model explains almost as much variation in SBP as does the model containing all three of the predictors and does so with one less variable included in the model.

5. a

	df	Sum of Squares	Mean Sum of Squares	F
Regression	3	25974	8658.0	80.871
Residual	21	2248.23	107.059	
	24	28222.23		

b $R^2 = \dfrac{28222.23 - 2248.23}{28222.23} = 0.920$

The r^2 value implies that there is a strong linear relationship between education resources and student performance.

9. a As temperature increases from 20 to 25, the oxygen consumption increases by 0.0394(25-20) = 0.197.

b As weight increases from 0.25 to 0.5, the oxygen consumption increases by 0.5921(0.50-0.25) = 0.148

c   **i** $R^2 = \dfrac{3.406 - 3.340}{3.406} = 0.019$

     ii $R^2 = \dfrac{3.406 - 0.632}{3.406} = 0.814$

     iii $R^2 = \dfrac{3.406 - 0.195}{3.406} = 0.943$

13. a $\hat{Y} = 6.874 - 0.004X_2 - 0.234X_3$

   b $\hat{Y} = 6.874 - 0.004(200) - 0.234(10) = 3.734$

   c $R^2 = \dfrac{26.583}{48.080} = 0.553$ The model explains over fifty percent of the variation in Yield (Y). The model has a limited ability to predict the yield for a company using 1989 ranking and P-E ratio as predictors.

Chapter 9

1. **a** **i** For the model $Y = \beta_0 + \beta_1 X_1 + E$ (X_1 = Age):
H_0: $\beta_1 = 0$ H_a: $\beta_1 \neq 0$
$F(X_1)_{1,30} = 45.18^*$, $p = 0.0001$.
At $\alpha = 0.05$, we would reject H_0 and conclude that $\beta_1 \neq 0$ with a p-value = 0.0001.
* F-test values will be taken from SAS output unless calculations are necessary.

   ii For the model $Y = \beta_0 + \beta_1 X_1 + \beta_2 X_2 + E$ (X_2 = SMK):
H_0: $\beta_1 = \beta_2 = 0$ H_a: at least one $\beta_i \neq 0$
$F(X_1, X_2)_{2,29} = 39.16$, $p = 0.0001$.
At $\alpha = 0.05$, we would reject H_0 and conclude that at least one $\beta_i \neq 0$ with a p-value = 0.0001.

   iii For the model $Y = \beta_0 + \beta_1 X_1 + \beta_2 X_2 + \beta_3 X_3 + E$ (X_3 = QUET):
H_0: $\beta_1 = \beta_2 = \beta_3 = 0$ H_a: at least one $\beta_i \neq 0$
$F(X_1, X_2, X_3)_{3,28} = 29.71$, $p = 0.0001$.
At $\alpha = 0.05$, we would reject H_0 and conclude that at least one $\beta_i \neq 0$ with a p-value = 0.0001.

b Using the criteria of favoring a more parsimonious model, one would choose the model tested in 1.a(i). The model contains only one predictor and the F-test demonstrates that the linear relationship between AGE and SBP in highly statistically significant. The answer given for Chapter 8,1(b) preferred the model containing SMK and AGE as predictors of SBP.

5 **a** **i** H_0: $\beta_1 = 0$ in the model $Y = \beta_0 + \beta_1 X_1 + E$. H_a: $\beta_1 \neq 0$

$$F(X_1)_{1,40} = \frac{46.236}{\frac{4567.383}{40}} = 0.405, \quad p > 0.25.$$

At $\alpha = 0.05$, we do not reject H_0 and conclude that $\beta_1 = 0$.

   ii H_0: $\beta_2 = 0$ in the model $Y = \beta_0 + \beta_2 X_2 + E$. H_a: $\beta_2 \neq 0$
$F(X_2)_{1,40} = 0.19$, $p > 0.25$.
At $\alpha = 0.05$, we do not reject H_0 and conclude that $\beta_2 = 0$.

   iii H_0: $\beta_3 = 0$ in the model $Y = \beta_0 + \beta_3 X_3 + E$. H_a: $\beta_3 \neq 0$
$F(X_3)_{1,40} = 7.58$, $p = 0.009$.
At $\alpha = 0.05$, we reject H_0 and conclude $\beta_3 \neq 0$ with a p-value = 0.009.

b H_0: $\beta_1 = \beta_2 = \beta_3 = 0$ in the model $Y = \beta_0 + \beta_1 X_1 + \beta_2 X_2 + \beta_3 X_3 + E$. H_a: at least one $\beta_i \neq 0$

$$F(X_1, X_2, X_3)_{3,38} = \frac{\frac{684.365 + 89.147 + 46.236}{3}}{\frac{3793.872}{38}} = 2.737, \quad 0.05 < p < 0.10.$$

At $\alpha = 0.05$, we do not reject H_0; X_1, X_2, X_3 taken together do not significantly help to predict Y.

c $Y = \beta_0 + \beta_1 X_1 + \beta_2 X_2 + \beta_3 X_3 + \beta_4 X_1 X_3 + \beta_5 X_2 X_3 + E$ equals full model.
$Y = \beta_0 + \beta_1 X_1 + \beta_2 X_2 + \beta_3 X_3 + E$ equals reduced model.
H_0: $\beta_4 = \beta_5 = 0$ in the model $Y = \beta_0 + \beta_1 X_1 + \beta_2 X_2 + \beta_3 X_3 + \beta_4 X_1 X_3 + \beta_5 X_2 X_3 + E$
H_a: at least one $\beta_i \neq 0$

$$F(X_1 X_3, X_2 X_3 | X_1, X_2, X_3)_{2,36} = \frac{\frac{48.690 + 26.13}{2}}{\frac{3719.052}{36}} = 0.362, \; p > 0.25.$$

At $\alpha = 0.05$, we do not reject H_0 and conclude that at least one $\beta_i = 0$. We may infer that the relationship of Y to X_1 and X_2 does not change when the value of X_3 changes. The resulting regression lines when X_3 changes value will be parallel.

d H_0: $\beta_3 = 0$ in the model $Y = \beta_0 + \beta_1 X_1 + \beta_2 X_2 + \beta_3 X_3 + E$. H_a: $\beta_3 \neq 0$

$$F(X_3 | X_1, X_2)_{1,38} = \frac{684.365}{\frac{3793.796}{38}} = 6.855, \; 0.01 < p < 0.025.$$

At $\alpha = 0.05$, we reject H_0 and conclude that $\beta_3 \neq 0$. The regression when X_3 is absent, is not coincident with the equation when X_3 is present. The results that X_3 is a significant predictor for Y, but interaction between X_3 and the other predictors is not significant.

e X_3 is associated with Y but the other two independent variables are not and should not be included in the model.

9.   **a** H_0: $\beta_1 = \beta_2 = 0$ in the model $Y = \beta_0 + \beta_1 X_1 + \beta_2 X_2 + E$. H_a: at least one $\beta_i \neq 0$
$F(X_1, X_2)_{2,4} = 20.03, \; p = 0.0082$.
At $\alpha = 0.05$, we would reject H_0 and conclude that at least one $\beta_i \neq 0$.

   b   **i** H_0: $\beta_1 = 0$ in the model $Y = \beta_0 + \beta_1 X_1 + E$. H_a: $\beta_1 \neq 0$
$F(X_1)_{1,5} = 40.06, \; p = 0.0032$.
At $\alpha = 0.05$, we would reject H_0 and conclude that $\beta_1 \neq 0$.

     ii H_0: $\beta_2 = 0$ in the model $Y = \beta_0 + \beta_1 X_1 + \beta_2 X_2 + E$. H_a: $\beta_2 \neq 0$
$F(X_2 | X_1)_{1,4} = 0.01, \; p = 0.9344$.
At $\alpha = 0.05$, we do not reject H_0 and conclude that at $\beta_2 = 0$.

   c   **i** H_0: $\beta_2 = 0$ in the model $Y = \beta_0 + \beta_2 X_2 + E$. H_a: $\beta_2 \neq 0$

$$F(X_2)_{1,5} = \frac{5733.321 - 1402.315}{\frac{572.393}{5}} = 37.832, \; \text{with } 0.001 < p < 0.005.$$

At $\alpha = 0.05$, we would reject H_0 and conclude that $\beta_1 \neq 0$.

     ii H_0: $\beta_1 = 0$ in the model $Y = \beta_0 + \beta_1 X_1 + \beta_2 X_2 + E$. H_a: $\beta_1 \neq 0$
$F(X_1 | X_2)_{1,4} = 9.80, \; p = 0.0352$.
At $\alpha = 0.05$, we reject H_0 and conclude that $\beta_1 \neq 0$.

d
Source	df	SS	MS	F	R^2
$X_1 \mid X_2$	1	1402.315	1402.315	9.8	0.909
$X_2 \mid X_1$	1	1.098	1.098	0.01	
Residual	4	572.393	143.098		
Total	6	6305.714			

e X_1 is the only necessary predictor. If one considered X_2 an extremely important variable, it may be considered a necessary predictor. However, the inclusion of X_1 in the model is necessary for X_2 to significantly add to the model. X_2's presence is not required for X_1 to be significant and therefore is not a necessary variable.

13. a H_0: $\beta_1 = \beta_2 = 0$ in the model $Y = \beta_0 + \beta_1 X_1 + \beta_2 X_2 + E$.
H_a: at least one $\beta_i \neq 0$, (X_1 = OWNCOST, X_2 = URBAN)
$F(X_1, X_2)_{2,23} = 8.521$, $p = 0.017$.
At $\alpha = 0.05$, we would reject H_0 and conclude that at least one $\beta_i \neq 0$.

b i H_0: $\beta_1 = 0$ in the model $Y = \beta_0 + \beta_1 X_1 + E$. H_a: $\beta_1 \neq 0$
$F(X_1)_{1,24} = 8.84$, $p = 0.0068$.
At $\alpha = 0.05$, we reject H_0 and conclude that $\beta_1 \neq 0$.

ii H_0: $\beta_2 = 0$ in the model $Y = \beta_0 + \beta_1 X_1 + \beta_2 X_2 + E$. H_a: $\beta_2 \neq 0$
$F(X_2 \mid X_1)_{1,23} = 8.21$, $p = 0.0088$.
At $\alpha = 0.05$, we reject H_0 and conclude that at $\beta_2 \neq 0$.

c i H_0: $\beta_2 = 0$ in the model $Y = \beta_0 + \beta_2 X_2 + E$. H_a: $\beta_2 \neq 0$

$$F(X_2) = \frac{255.779 - 66.334}{\frac{345.183}{24}} = 13.17 \text{ with 1 and 24 df, } 0.001 < p < 0.005.$$

At $\alpha = 0.05$, we reject H_0 and conclude that at $\beta_2 \neq 0$.

ii H_0: $\beta_1 = 0$ in the model $Y = \beta_0 + \beta_1 X_1 + \beta_2 X_2 + E$. H_a: $\beta_1 \neq 0$
$F(X_1 \mid X_2)_{1,23} = 4.42$, $p = 0.0467$.
At $\alpha = 0.05$, we reject H_0 and conclude that at $\beta_1 \neq 0$.

d
Source	df	SS	MS	F	R^2
$X_1 \mid X_2$	1	66.334	66.334	4.42	0.4256
$X_3 \mid X_2$	1	123.158	123.158	8.21	
Residual	23	345.183	15.008		
Total	25	600.962			

e Both predictors are necessary. Each variable is significant regardless of the order entered in the model and both should be included.

Chapter 10

1. **a** Age with a r value of 0.7752.

b **i** $r_{SBP,\ SMK|AGE} = \sqrt{0.323} = 0.568$. The value, 0.323, is taken from the SAS Squared Partial Correlation Type I provided for Problem 1.

ii $r_{SBP,\ QUET|AGE} = \sqrt{0.101} = 0.318$

c $H_0 : \rho_{SBP,\ SMK\ |\ AGE} = 0 \qquad H_a : \rho_{SBP,\ SMK|AGE} \neq 0$

$$F(SMK\ |\ AGE)_{1,29} = \frac{4689.684 - 3861.630}{\frac{1736.285}{29}} = 13.83\ ,\ p < 0.001.$$

Note: Instead of calculating this partial F statistic using the formula above, we could have determined it by calculating the square of the appropriate partial t-statistic (3.719) on the SAS output.

At $\alpha = 0.05$, we reject H_0 and conclude that SMK added to a model already containing AGE explains a significant amount of variation in SBP.

d $H_0 : \rho_{SBP,\ QUET\ |\ AGE,\ SMK} = 0 \qquad H_a : \rho_{SBP,\ QUET\ |\ AGE,\ SMK} \neq 0$

$T_{28} = 1.91$, $p = 0.066$.

At $\alpha = 0.05$, we do not reject H_0 and conclude that QUET added to a model already containing AGE and SMK does not explain a significant amount of variation in SBP.

e Based on the results for a-d we find that the following variables ranked in order of their significance in explaining the variation in SBP: 1) AGE, 2) SMK, 3) QUET. QUET may be considered relatively unimportant since the two other predictors, AGE and SMK explain most of the variation in SBP, with the proportions of explained variation both being significant at $\alpha=0.05$, while QUET does not.

f $r^2_{SBP(QUET,\ SMK)\ |\ AGE} = \frac{4889.826 - 3861.630}{2564.338} = 0.401$

$H_0 : \rho_{SBP\ (QUET,\ SMK)\ |\ AGE} = 0 \qquad H_a : \rho_{SBP\ (QUET,\ SMK)\ |\ AGE} \neq 0$

$$F(QUET,\ SMK\ |\ AGE)_{2,28} = \frac{\frac{4889.826 - 3861.630}{2}}{\frac{1536.143}{28}} = 9.371\ ,\ p < 0.001.$$

The highly significant p-value suggests that both SMK and QUET are important variables, but there is room for debate since the increase in r^2 going from model 1, with only AGE (0.601), to model 3, with all 3 variables (0.761) is small (0.160).

5. **a** **i** $r^2_{YX_1 | X_3} = \dfrac{SS\ reg\ (X_1 | X_3)}{SS\ residual\ (X_3)} = \dfrac{35.632}{1855.202 - 1387.6} = 0.076$

ii $r^2_{YX_2 | X_3} = \dfrac{100.26}{1855.202 - 1387.6} = 0.214$

iii $r^2_{YX_2 | X_3} = \dfrac{(r_{YX_2} - r_{YX_3} * r_{X_2 X_3})^2}{(1 - r^2_{YX_3})(1 - r^2_{X_2 X_3})} = \dfrac{(0.8398 - 0.8648 * 0.8154)^2}{(1 - 0.8648^2)(1 - 0.8154^2)} = 0.215$

The computations are nearly equivalent with the difference being due round-off error.

b X_2 should be considered next for entry into the model because X_2 has a higher partial correlation than does X_1.

c $H_0 : \rho_{YX_2 | X_3} = 0 \qquad H_a : \rho_{YX_2 | X_3} \neq 0$

$T_{17} = 0.463 \dfrac{\sqrt{20 - 1 - 2}}{\sqrt{1 - 0.463^2}} = 2.154 \ , 0.02 < p < 0.05.$

At $\alpha = 0.05$ we reject H_0 and conclude that X_2 added to a model already containing X_3 does explain a significant amount of variation in Y.

d $r^2_{YX_1 | X_2, X_3} = 0.082$

$H_0 : \rho_{YX_1 | X_2, X_3} = 0 \qquad H_a : \rho_{YX_1 | X_2, X_3} \neq 0$

$T_{16} = 1.199, \ p = 0.248.$

At $\alpha = 0.05$ we do not reject H_0 and conclude that X_1 added to a model already containing X_2 and X_3 does not explain a significant amount of variation in Y.

e $r^2_{Y(X_1, X_2) | X_3} = \dfrac{1518.145 - 1387.6}{1855.202 - 1387.6} = 0.279$

$H_0 : \rho_{Y(X_1, X_2) | X_3} = 0 \qquad H_a : \rho_{Y(X_1, X_2) | X_3} \neq 0$

$F(X_1, X_2 | X_3)_{2,16} = \dfrac{\dfrac{1518.145 - 1387.6}{2}}{\dfrac{337.057}{16}} = 3.098 \ , 0.05 < p < 0.10.$

At $\alpha = 0.05$ we do not reject H_0 and conclude that X_1 and X_2 added to a model already containing X_3 does not explain a significant amount of variation in Y.

f Based on the above results, only X_3 should be included in the model at $\alpha = 0.05$.

9. a i $H_0 : \rho_{YX_1} = 0 \quad H_a : \rho_{YX_1} \neq 0$

   $F(X_1)_{1,45} = 0.89, \; p = 0.35.$
   At $\alpha = 0.05$ we do not reject H_0 and conclude that X_1 alone does not explain a significant amount of variation in Y.

   ii $H_0 : \rho_{YX_2} = 0 \quad H_a : \rho_{YX_2} \neq 0$

   $F(X_2)_{1,45} = 197.58, \; p = 0.0001.$
   At $\alpha = 0.05$ we reject H_0 and conclude that X_2 alone explains a significant amount of variation in Y.

   b i $H_0 : \rho_{YX_1 | X_2} = 0 \quad H_a : \rho_{YX_1 | X_2} \neq 0$

   $$F(X_1 | X_2)_{1,44} = \frac{3.211 - 2.774}{\frac{0.195}{44}} = 98.605, \; p < 0.0001.$$

   At $\alpha = 0.05$ we reject H_0 and conclude that X_1 added to a model already containing X_2 does explain a significant amount of variation in Y.

   ii $H_0 : \rho_{YX_2 | X_1} = 0 \quad H_a : \rho_{YX_2 | X_1} \neq 0$

   $$F(X_2 | X_1)_{1,44} = \frac{3.211 - 0.066}{\frac{0.195}{44}} = 709.641, \; p < 0.0001.$$

   At $\alpha = 0.05$ we reject H_0 and conclude that X_2 added to a model already containing X_1 does explain a significant amount of variation in Y.

   c Both X_1 and X_2 should be included in the model with X_2 being the more important than X_1.

13. a $r^2_{Y | X_1, X_2} = 0.909$ b $r_{YX_2 | X_1} = \sqrt{0.002} = -0.045$ c $r_{YX_1 | X_2} = \sqrt{0.710} = 0.84$

   d $H_0 : \rho_{YX_2 | X_1} = 0 \quad H_a : \rho_{YX_2 | X_1} \neq 0$

   $$T_4 = -0.045 \frac{\sqrt{7-3}}{\sqrt{1-0.002}} = -0.09, \; p > 0.90.$$

   At $\alpha = 0.05$ we do not reject H_0 and conclude that X_2 added to a model already containing X_1 does not explain a significant amount of variation in Y. The value in the SAS output is virtually identical to the value calculated above.

e $H_0 : \rho_{YX_1 | X_2} = 0 \quad H_a : \rho_{YX_1 | X_2} \neq 0$

$$T_4 = 0.843 \frac{\sqrt{7-3}}{\sqrt{1-0.710}} = 3.131 \; , 0.02 < p < 0.05.$$

At $\alpha = 0.05$ we reject H_0 and conclude that X_1 added to a model already containing X_2 does explain a significant amount of variation in Y. The value in the SAS output is virtually identical to the value calculated above.

f X_1 should be included in the model while X_2 should not be included. X_1 is clearly a more important predictor of Y than X_1.

17. a $r^2_{Y | X_1, X_2} = 0.426$ **b** $r_{YX_2 | X_1} = \sqrt{0.263} = -0.513$ **c** $r_{YX_1 | X_2} = \sqrt{0.161} = -0.40$

d $H_0 : \rho_{YX_2 | X_1} = 0 \quad H_a : \rho_{YX_2 | X_1} \neq 0$

$$T_{23} = -0.513 \frac{\sqrt{26-3}}{\sqrt{1-0.263}} = -2.865 \; , 0.001 < p < 0.01.$$

At $\alpha = 0.05$ we reject H_0 and conclude that X_2 added to a model already containing X_1 does explain a significant amount of variation in Y. The value in the SAS output given for Chap 9 Quest 13 is virtually identical to the value calculated above.

e $H_0 : \rho_{YX_1 | X_2} = 0 \quad H_a : \rho_{YX_1 | X_2} \neq 0$

$$T_{23} = -0.401 \frac{\sqrt{26-3}}{\sqrt{1-0.161}} = -2.1 \; , 0.02 < p < 0.05.$$

At $\alpha = 0.05$ we reject H_0 and conclude that X_1 added to a model already containing X_2 does explain a significant amount of variation in Y. The value in the SAS output given for Chap 9 Quest 13 is virtually identical to the value calculated above.

f   Both variables should be included in the model with X_2 being the more important predictor of Y.

Chapter 11

1. **a** WGT = $\beta_0 + \beta_1$HGT + β_2AGE + β_3AGE^2 + E
   b Since we are interested in the relationship between HGT and WGT, we will focus on $\hat{\beta}_1$ as the measure of association. The following table is useful to assess potential confounding due to AGE and/or AGE^2:

Independent Variables In Model	$\hat{\beta}_1$	95 % CI for β_1
HGT, AGE, AGE^2	0.72	(0.097, 1.350)
HGT, AGE	0.72	(0.141, 1.303)
HGT, AGE^2	0.73	(0.139, 1.313)
HGT	1.07	(0.540, 1.604)

   Note that $\hat{\beta}_1$ does not change when either AGE or AGE^2 are removed from the model. However, $\hat{\beta}_1$ changes "significantly" when both AGE and AGE^2 are removed from the model. Thus, there is confounding due to AGE and AGE^2.

   c AGE^2 can be dropped from the model because $\hat{\beta}_1$ does not change significantly.

   d AGE^2 should not be retained in the model because the 95% C.I. for β_1 is narrower when AGE^2 is absent from the model.

   e Considering the change in $\hat{\beta}_1$ and the width of the 95% C.I., the final model should be WGT = $\beta_0 + \beta_1$HGT + β_2AGE + E.

   f Revise the model as WGT = $\beta_0 + \beta_1$HGT + β_2AGE + β_3AGE^2 + β_4HGT*AGE + β_5HGT*AGE^2 + E.

   g We would test for interaction by performing a multiple-partial F test for H_0: $\beta_4 = \beta_5 = 0$. If this test is significant, then perform separate partial F tests to assess H_0: $\beta_4 = 0$ and H_0: $\beta_5 = 0$.

3. **a** There is no confounding due to X_2 because $\hat{\beta}_1$ does not change when X_2 is removed from the model.

   b $r_{YX_1} = 0.265 \quad r_{YX_1|X_2} = \sqrt{0.5} = 0.707$

   Since the two correlation coefficients are significantly different, we conclude that confounding exists.

   c We note from parts (a) and (b) that the conclusions for confounding depend on the definition of confounding.

   d Since H_0: $\beta_2 = 0$ is rejected (p=0.0005), we conclude that confounding exists which is contradictory to part (a).

5. **a** **i** H_0: $\beta_1 = 0 \quad H_a$: $\beta_1 \neq 0$

   $F(X_1)_{1,40} = 0.4 \quad p = 0.528$
   At $\alpha = 0.05$, we do not reject H_0.

ii $H_0: \beta_2 = 0$ $H_a: \beta_2 \neq 0$

   $F(X_2)_{1,40} = 0.19$ $p = 0.669$
   At $\alpha = 0.05$, we do not reject H_0.

iii $H_0: \beta_3 = 0$ $H_a: \beta_3 \neq 0$

   $F(X_3)_{1,40} = 7.58$ $p = 0.009$
   At $\alpha = 0.05$, we reject H_0.

b $H_0: \beta_1 = \beta_2 = \beta_3 = 0$ $H_a:$ at least one $\beta_i \neq 0$

$$F(X_1, X_2, X_3)_{3,38} = \frac{\frac{819.748}{3}}{\frac{4613.619 - 819.748}{38}} = 2.737 \text{ with } 0.05 < p < 0.1.$$

At $\alpha = 0.05$, we do not reject H_0.

c $H_0: \rho_{Y(X_1X_3,\, X_2X_3)|\, X_1,\, X_2,\, X_3} = 0$ $H_a: \rho_{Y(X_1X_3,\, X_2X_3)|\, X_1,\, X_2,\, X_3} \neq 0$

$$F(X_1X_3, X_2X_3 | X_1, X_2, X_3)_{2,36} = \frac{\frac{74.820}{2}}{\frac{3719.052}{36}} = 0.362 \, , p > 0.25.$$

Since this test is not significant, the only regression coefficient that changes (depending upon whether $X_3 = 0$ or 1) is the intercept. Thus, the two estimated regression equations are parallel to each other and separated by a distance equal to $\hat{\beta}_3$.

d $H_0: \rho_{YX_3|\, X_1,\, X_2} = 0$ $H_a: \rho_{YX_3|\, X_1,\, X_2} \neq 0$

$$F(X_3 | X_1, X_2)_{1,38} = \frac{684.365}{\frac{3793.871}{38}} = 6.855 \, , p = 0.014.$$

At $\alpha = 0.05$, we reject H_0 which implies that the regression equation when X_3 is absent is not coincident with the equation when X_3 is present.

e First fit the full model
$$Y = \beta_0 + \beta_1 X_1 + \beta_2 X_2 + \beta_3 X_3 + E. \quad (1)$$
Next, fit the reduced models
$$Y = \beta_0 + \beta_1 X_1 + \beta_3 X_3 + E. \quad (2)$$
$$Y = \beta_0 + \beta_2 X_2 + \beta_3 X_3 + E. \quad (3)$$
$$Y = \beta_0 + \beta_3 X_3 + E. \quad (4)$$
We assess confounding by noting how $\hat{\beta}_3$ changes for the different models. In particular, if $\hat{\beta}_3$ from models (2), (3), or (4) is different from $\hat{\beta}_3$ of model (1), then X_1, X_2, or X_1 and X_2, respectively, are confounders. To assess precision, we note how the $100(1-\alpha)\%$ C.I.'s for $\hat{\beta}_3$ change. We only eliminate <u>potential</u> confounders from the model if the width of the C.I. for $\hat{\beta}_3$ does not widen significantly.

f From the information provided, we can assess the confounding effects of X_1 or X_2 alone with respect to X_3 but not for X_1 and X_2 taken together.

7.
a No, there is no meaningful change in the estimate for $\hat{\beta}_1$ when adding X_2 to the model.
b No, the confidence interval for $\hat{\beta}_1$ is narrower when only X_1 is in the model.
c No, there is not evidence suggesting that including X_2 to the model improves the precision and/or the validity of the estimated relationship between X_1 and Y.

9.
a Let OWNCOST = X_1 and INCOME = X_2 in the model $Y = \beta_0 + \beta_1 X_1 + \beta_2 X_2 + E$
H_0: $\beta_1 = \beta_2 = 0$ H_a: at least one β_i does not equal 0.
$F(X_1, X_2)_{2,23} = 6.38$, $p = 0.006$.
At $\alpha = 0.05$, we reject H_0 and conclude that X_1 and X_2 taken together do significantly help predict Y.
b H_0: $\beta_1 = 0$ in the model $Y = \beta_0 + \beta_1 X_1 + \beta_2 X_2 + E$. H_a: $\beta_1 \neq 0$
$F(X_1 | X_2)_{1,23} = 11.47$, $p = 0.003$.
At $\alpha = 0.05$, we reject H_0 and conclude that X_1 added to a model already containing X_2 does significantly help predict Y.
c H_0: $\beta_2 = 0$ in the model $Y = \beta_0 + \beta_1 X_1 + \beta_2 X_2 + E$. H_a: $\beta_2 \neq 0$
$F(X_2 | X_1)_{1,23} = 4.87$, $p = 0.038$.
At $\alpha = 0.05$, we reject H_0 and conclude that X_2 does contribute significantly to the model already containing X_1.
d Including X_2 does meaningfully change $\hat{\beta}_1$ and should therefore be included in the model as a confounder assuming there is no interaction between X_1 and X_2.

Chapter 12

Many of the regression models in this chapter have been estimated in previous chapters. Please refer to the solutions in those earlier chapters for estimates of the models.

1. **a** $Y = \beta_0 + \beta_1 (AGE) + E$, in which Y denotes dry weight.
   b Use output provided for Question 1.
   c Use output provided for Question 1.
   d The largest (absolute value) jackknife residual is $r_{(-11)} = 3.568$. This value is then compared to a two-sided (Bonferroni corrected) critical region of the t-distribution, $t_{8, (0.025/11)} = 3.9$. Thus this jackknife residual is not significant.
   e The jackknife residual in part (d) should not be overlooked (even though it was not significant) since it is nearly three times bigger than the next largest jackknife residual, $r_{(-1)} = 1.315$. However, the primary problem is illustrated by the plot of the jackknife residuals vs. AGE. The systematic pattern of the residuals suggests that the straight line regression model is not appropriate.
   f The jackknife residuals follow an exact t-distribution. For samples less than 30, one should expect heavier tails than for a normal curve. As sample size increases, i.e. above 30, the distribution should reflect sampling from a standard normal distribution.

6. **a** $Y = \beta_0 + \beta_1 (AGE) + E$, in which Y denotes average total sleep time.
   b

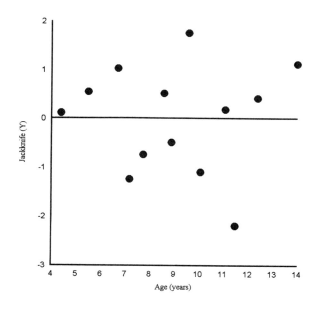

c

```
Stem Leaf        #     Boxplot
  1 018          3        |
  0 12456        5     *-----*
 -0 75           2     +-----+
 -1 21           2        |
 -2 2            1        |
 ----+----+----+----+
```

d The largest (absolute value) jackknife residual is $r_{(-5)} = 2.190$. This value is then compared to a two-sided (Bonferroni corrected) critical region of the t-distribution, $t_{10, (0.025/13)} = 3.740$. Thus, this jackknife residual is not significant.

e In general the residuals appear to be in good shape. None appear to be too extreme and no pattern is easily recognized.

13. a $Y = \beta_0 + \beta_1 (\text{BODY WEIGHT}) + E$, in which Y denotes latency to seizure.

b

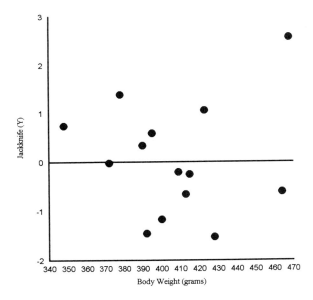

c

```
Stem Leaf        #     Boxplot
  2 6            1        |
  1 14           2        |
  0 367          3     +-----+
 -0 76320        5     *-----*
 -1 552          3        |
 ----+----+----+----+
```

- **d** The largest (absolute value) jackknife residual is $r_{(-14)}= 2.558$. This value is then compared to a two-sided (Bonferroni corrected) critical region of the t-distribution, $t_{11, (0.025/14)} = 3.689$. Thus, this jackknife residual is not significant.
- **e** The plots of the jackknife residuals do not suggest any troublesome observations other than the one discussed in part (d).

20. a $Y = \beta_0 + \beta_1 (\text{HEIGHT}) + \beta_2 (\text{WEIGHT}) + \beta_3 (\text{AGE}) + \beta_4 (\text{FEMALE}) + E$, in which Y denotes forced expiratory volume in one second.

b

Variable-added-last tests

Variable	SS	F
INTERCEPT	0.013	0.016
HEIGHT	0.312	0.395
WEIGHT	0.478	0.603
AGE	0.047	0.059
FEMALE	1.605	2.028

The critical value is $F_{1, 14, 0.95} = 4.60$ and hence none of the predictors significantly explain the variation in FEV_1.

- **c** See output provided in textbook for Question 20.
- **d** See Problem 19 (d).
- **e** See output in textbook for Question 20.
- **f** See output provided in textbook for Question 20.
- **g** See SAS output provided in textbook for Question 20. There are no obvious problems apparent from the residuals and leverage values.
- **h** Collinearity of the predictors does not appear to be a problem. However, the variance proportions corresponding to the large condition number for the scaled cross products matrix suggest that height and the intercept are moderately collinear.

25. a $Y = \beta_0 + \beta_1 (\text{ADVERTISING}) + E$, in which Y denotes sales.
- **b** See output provided in textbook for Question 25.
- **c** See output provided in textbook for Question 25.
- **d** The largest studentized residual (absolute value) = 1.527 which is no cause for alarm.
- **e** The plot of the jackknife residuals versus the predictor does not suggest any troublesome problems. Due to the number of observations being so small, it is hard to assess the frequency histogram and schematic plot. However, they do not show any obvious problems.

Chapter 13

1. **a** Plots of Y on X and ln Y on X are shown below. The least squares line and the least squares parabola are also shown.

Q.1.a (i)   **Q.1.a (ii)**

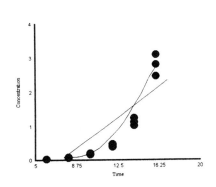 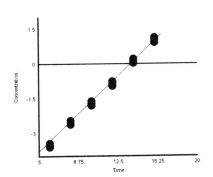

b From the computer output we find:

    (1) Degree 1: $\hat{Y} = -1.932 + 0.246X$

    (2) Degree 2: $\hat{Y} = 3.172 - 0.781X + 0.047X^2$

    (3) ln Y on X: $\ln\hat{Y} = -6.21 + 0.451X$

    (4) The above fitted equations are plotted on the graphs presented for 1(a).

c

Source	df	SS	MS	F
Regression	1	12.705	12.705	43.69
Lack of fit	4	4.419	1.105	57.03
Residual	16	4.651	0.2908	
Pure Error	12	0.232	0.0194	
Total	17	17.357		

d

Source	df	SS	MS	F
Degree 1 (X)	1	12.705	12.705	43.69
Regression	2	16.61	8.305	
Degree 2 ($X^2 \mid X$)	1	3.905	3.905	78.46
Lack of fit	3	0.514	0.171	8.85
Residual	15	0.746	0.0497	
Pure Error	12	0.232	0.0194	
Total	17	17.357		

e $r^2_{XY} = 0.732$; $r^2(quadratic) = 0.957$

f Test for significance of straight line regression of Y on X
H_0: The straight line regression is not significant.

$$F_{1,16} = \frac{MS\ Reg(X)}{MS\ Res(X)} = \frac{12.705}{0.2908} = 43.69 \text{, with } p < 0.001.$$

At $\alpha = 0.05$ we reject H_0 and conclude that the straight line regression is significant.
Test for adequacy of straight line model
H_0: The straight line model is adequate.

$$F_{4,12} = \frac{MS\ l.o.f.\ (X)}{MS\ P.E.\ (X)} = \frac{1.105}{0.0194} = 56.959 \text{, with } p < 0.001.$$

At $\alpha = 0.05$ we reject H_0 and conclude that the straight line model is not adequate.

g Test for significance of quadratic regression
H_0: The quadratic regression is not significant.

$$F_{2,15} = \frac{MS\ Reg(X, X^2)}{MS\ Res(X, X^2)} = \frac{8.305}{0.0498} = 166.77 \text{, with } p = 0.0001.$$

At $\alpha = 0.05$ we reject H_0 and conclude that the quadratic regression is significant.
Test for addition of X^2 term
H_0: The addition of X^2 to a model already containing X is not significant.

$$F(X^2 \mid X)_{1,15} = \frac{3.905}{0.0498} = 78.41 \text{, with } p = 0.0001.$$

At $\alpha = 0.05$ we reject H_0 and conclude that the addition of X^2 is significant.
Test for adequacy of quadratic model
H_0: The quadratic model is adequate.

$$F_{3,12} = \frac{MS\ l.o.f.\ (X, X^2)}{MS\ P.E.\ (X, X^2)} = \frac{0.1715}{0.0194} = 8.84 \text{, with } p = 0.002.$$

At $\alpha = 0.05$ we reject H_0 and conclude that the quadratic model is not adequate.

h Test for significance of straight line regression of ln Y on X
H$_0$: The straight line regression is not significant.

$$F_{1,16} = \frac{MS\ Reg(X)}{MS\ Res(X)} = \frac{42.746}{0.009994} = 4277.167 \text{ , with } p = 0.0001.$$

At $\alpha = 0.05$ we reject H$_0$ and conclude that the straight line regression is significant.
Test for adequacy of straight line model of ln Y on X
H$_0$: The straight line model is adequate.

$$F_{4,12} = \frac{MS\ l.o.f.\ (X)}{MS\ P.E.\ (X)} = \frac{0.009146}{0.010206} = 0.896 \text{ , with } p = 0.471.$$

At $\alpha = 0.05$ we do not reject H$_0$ and conclude that the straight line model is adequate.

i R^2 (straight line regression of ln Y on X) = 0.9965
R^2 (quadratic regression of Y on X) = 0.957
A comparison of the above two R^2 shows that the straight line fit of ln Y on X provides a better fit to the data than the quadratic model of Y on X.

j (1) Homoscedasticity assumption appears to be much more reasonable when using ln Y on X than when using Y on X.
(2) The straight line regression of ln Y on X is preferred since it results in a higher R^2, the model is accurate, satisfies the assumption of homoscedasticity, and provides a better graphical fit.

k The independence assumption is violated.

5. a

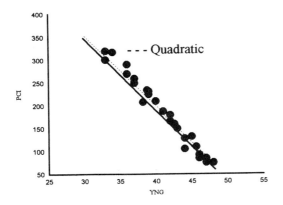

b Test for significance of straight line regression
H_0: The straight line regression is not significant.

$$F_{1,24} = \frac{MS\ Reg(X)}{MS\ Res(X)} = \frac{153784.8}{157.238} = 978.04 \text{ , with } p < 0.001.$$

At $\alpha = 0.05$ we reject H_0 and conclude that the straight line regression is significant.
Test for adequacy of straight line model
H_0: The straight line model is adequate.

$$F_{16,8} = \frac{MS\ l.o.f.\ (X)}{MS\ P.E.\ (X)} = \frac{178.888}{113.9} = 1.57 \text{ , with } p > 0.25.$$

At $\alpha = 0.05$ we do not reject H_0 and conclude that the straight line model is adequate.

c Test for addition of X^2 to the model
H_0: The addition of X^2 is not significant.
$F(X^2 | X)_{1,23} = 88.3 / 160.235 = 0.55$, with $p > 0.25$.
At $\alpha = 0.05$ we do not reject H_0.

d The straight line model is most appropriate.

9. a Fit the model VOC_SIZE = $\beta_0 + \beta_1 (AGE^*) + \beta_2 (AGE^*)^2 + \beta_3 (AGE^*)^3 + E$ in which $AGE^* = AGE - 2.867$.

$\hat{\beta}_0 = 741.84 \qquad \hat{\beta}_1 = 645.60 \qquad \hat{\beta}_2 = 70.43 \qquad \hat{\beta}_3 = -31.18$

b Using variables-added-in-order tests, the best model includes AGE^*, $(AGE^*)^2$, and $(AGE^*)^3$.

c Using variables-added-last tests, the best model includes AGE^*, $(AGE^*)^2$, and $(AGE^*)^3$, which is the same model as in part (b).

d

Collinearity Diagnostics

Variable	Eigenvalue	Condition Index	Variance Proportions		
			AGE*	(AGE*)²	(AGE*)³
1	2.35	1.00	0.02	0.05	0.02
2	0.59	2.00	0.07	0.54	0.01
3	0.06	6.05	0.90	0.41	0.98

Predictor Correlations

	AGE*	(AGE*)²	(AGE*)³
AGE*	1.00	0.44	0.90
(AGE*)²		1.00	0.66
(AGE*)³			1.00

The only large predictor correlation is between AGE^* and $(AGE^*)^3$. The largest condition index, i.e. $CI_3 = 6.05$, suggests that the centered data do not have any serious collinearity problems.

e

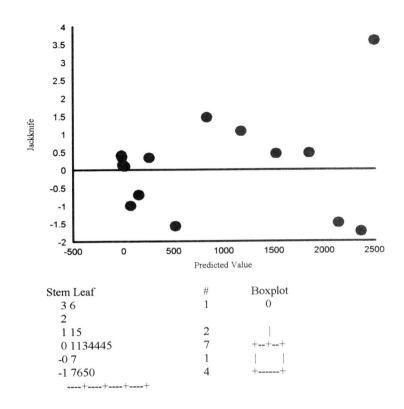

```
Stem Leaf            #     Boxplot
  3 6                1        0
  2
  1 15               2        |
  0 1134445          7      +--+--+
 -0 7                1      |    |
 -1 7650             4      +------+
  ----+----+----+----+
```

The comments for problem 8 also apply here.

f The estimated regression coefficients are different than in problem 8, but the best model includes the linear, quadratic, and cubic terms as in problem 8. Also, the sums of squares for the variables-added-in-order test are the same as in problem 8. The centering of AGE greatly reduced the previous collinearity problems. Note that centering does not affect the residual diagnostics.

13. a The orthogonal polynomial codings are:

```
0.129099  0.198860  0.101918  0.008216
0.129099  0.198860  0.101918  0.008216
0.129099  0.198860  0.101918  0.008216
0.129099  0.198860  0.101918  0.008216
0.129099  0.198860  0.101918  0.008216
0.129099 -0.122701  0.089127 -0.164316
0.129099 -0.122701  0.089127 -0.164316
0.129099 -0.122701  0.089127 -0.164316
0.129099 -0.122701  0.089127 -0.164316
0.129099 -0.122701  0.089127 -0.164316
0.129099 -0.122701  0.089127 -0.164316
0.129099 -0.122701  0.089127 -0.164316
0.129099 -0.122701  0.089127 -0.164316
0.129099 -0.122701  0.089127 -0.164316
0.129099 -0.122701  0.089127 -0.164316
0.129099 -0.122701  0.089127 -0.164316
0.129099 -0.122701  0.089127 -0.164316
0.129099 -0.122701  0.089127 -0.164316
0.129099 -0.105777  0.027129  0.195129
0.129099 -0.105777  0.027129  0.195129
0.129099 -0.105777  0.027129  0.195129
0.129099 -0.105777  0.027129  0.195129
0.129099 -0.105777  0.027129  0.195129
0.129099 -0.105777  0.027129  0.195129
0.129099 -0.105777  0.027129  0.195129
0.129099 -0.105777  0.027129  0.195129
0.129099 -0.105777  0.027129  0.195129
0.129099 -0.105777  0.027129  0.195129
0.129099  0.198860  0.101918  0.008216
0.129099  0.198860  0.101918  0.008216
0.129099  0.198860  0.101918  0.008216
0.129099  0.198860  0.101918  0.008216
0.129099  0.198860  0.101918  0.008216
0.129099  0.198860  0.101918  0.008216
0.129099  0.198860  0.101918  0.008216
0.129099  0.198860  0.101918  0.008216
0.129099  0.198860  0.101918  0.008216
0.129099  0.029617 -0.218174 -0.039026
0.129099  0.029617 -0.218174 -0.039026
0.129099  0.029617 -0.218174 -0.039026
0.129099  0.029617 -0.218174 -0.039026
0.129099  0.029617 -0.218174 -0.039026
0.129099  0.029617 -0.218174 -0.039026
0.129099  0.029617 -0.218174 -0.039026
0.129099  0.029617 -0.218174 -0.039026

cont
0.129099  0.029617 -0.218174 -0.039026
0.129099  0.029617 -0.218174 -0.039026
0.129099  0.029617 -0.218174 -0.039026
0.129099  0.029617 -0.218174 -0.039026
0.129099  0.029617 -0.218174 -0.039026
0.129099  0.029617 -0.218174 -0.039026
0.129099  0.029617 -0.218174 -0.039026
0.129099 -0.105777  0.027129  0.195129
0.129099 -0.105777  0.027129  0.195129
0.129099 -0.105777  0.027129  0.195129
0.129099 -0.105777  0.027129  0.195129
0.129099 -0.105777  0.027129  0.195129
```

The estimated equation is:

$\hat{Y} = 10.64 + 94.42(\text{LIN\_TOL}) + 14.59(\text{QUAD\_TOL}) - 0.73(\text{CUB\_TOL})$, in which LN_TOL, QUAD_TOL, and COL_TOL denote the centered PPM_TOLU orthogonal polynomial coefficients.

b Using the variable-added-in-order tests, the best model includes LIN_TOL and QUAD_TOL.

c Using the variable-added-last tests, the best model includes LIN_TOL and QUAD_TOL, which is the same model as in part (b). Note that these tests are exactly the same.

d

Collinearity Diagnostics

		Condition	Variance Proportions		
Variable	Eigenvalue	Index	LIN_TOL	QUAD_TOL	CUB_TOL
1	1.00	1.00	0.00	0.90	0.15
2	1.00	1.00	0.77	0.02	0.16
3	1.00	1.00	0.23	0.08	0.69

Predictor Correlations

	LIN_TOL	QUAD_TOL	CUB_TOL
LIN_TOL	1.00	0.00	0.00
QUAD_TOL		1.00	0.00
CUB_TOL			1.00

The orthogonal polynomials are uncorrelated with each other which implies that any collinearities are eliminated as shown by the condition indices.

e

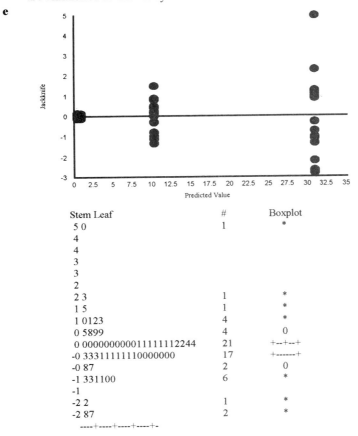

```
Stem Leaf                               #     Boxplot
  5 0                                   1       *
  4
  4
  3
  3
  2
  2 3                                   1       *
  1 5                                   1       *
  1 0123                                4       *
  0 5899                                4       0
  0 000000000011111112244              21     +--+--+
 -0 33311111110000000                  17     +------+
 -0 87                                  2       0
 -1 331100                              6       *
 -1
 -2 2                                   1       *
 -2 87                                  2       *
    ----+----+----+----+-
```

The residual plots clearly suggest that the variances increase as the predicted values increase. This problem is also highlighted by the detached residuals which appear on the schematic plot.

Chapter 14

1. a For smokers: $\hat{Y} = 79.225 + 20.118X$

For nonsmokers: $\hat{Y} = 49.312 + 26.303X$

b $H_0: \beta_{1SMK} = \beta_{1\overline{SMK}}$ $\qquad$ $H_a: \beta_{1SMK} < \beta_{1\overline{SMK}}$

Values obtained from the output provided for Question 1.

	n	$\hat{\beta}_0$	$\hat{\beta}_1$	$\overline{X}$	$\overline{Y}$	S_X^2	$S_{Y|X}^2$
Nonsmokers	15	49.312	26.303	3.478	140.8	0.176	48.274
Smokers	17	79.255	20.118	3.408	147.824	0.322	107.620

From these values, the pooled estimate of σ^2 is obtained as:

$$S_{P,Y|X}^2 = \frac{(n_{\overline{S}} - 2)S_{Y|X_{\overline{S}}}^2 + (n_S - 2)S_{Y|X_S}^2}{n_{\overline{S}} + n_S - 4} = \frac{(13*48.274) + (15*107.62)}{28} = 80.067$$

$$S_{(\hat{\beta}_{1\overline{SMK}} - \hat{\beta}_{1SMK})}^2 = S_{P,Y|X}^2\left[\frac{1}{(n_{\overline{S}} - 1)S_{X_{\overline{S}}}^2} + \frac{1}{(n_S - 1)S_{X_S}^2}\right] = 80.067\left[\frac{1}{14(0.176)} + \frac{1}{16(0.322)}\right] = 80.067(0.6) = 48.04$$

The test statistic for $H_0: \beta_{1SMK} = \beta_{1\overline{SMK}}$ equals:

$$T_{28} = \frac{\hat{\beta}_{1\overline{SMK}} - \hat{\beta}_{1SMK}}{S(\hat{\beta}_{1\overline{S}} - \hat{\beta}_{1S})} = \frac{26.303 - 20.118}{\sqrt{48.04}} = 0.892 \text{, with } 0.15 < p < 0.25.$$

At $\alpha = 0.05$ we do not reject H_0 that the slopes for smokers and nonsmokers are the same.

c $H_0: \beta_{0SMK} = \beta_{0\overline{SMK}}$ $\qquad$ $H_a: \beta_{0SMK} \neq \beta_{0\overline{SMK}}$

$$S_{(\hat{\beta}_{0\overline{SMK}} - \hat{\beta}_{0SMK})}^2 = S_{P,Y|X}^2\left[\frac{1}{n_{\overline{S}}} + \frac{1}{n_S} + \frac{\overline{X}_{\overline{S}}^2}{(n_{\overline{S}} - 1)S_{X_{\overline{S}}}^2} + \frac{\overline{X}_S^2}{(n_S - 1)S_{X_S}^2}\right]$$

$$= 80.067\left[\frac{1}{15} + \frac{1}{17} + \frac{(3.478)^2}{(14)(0.176)} + \frac{(3.408)^2}{(16)(0.322)}\right] = 583.528$$

The test statistic for $H_0: \beta_{0SMK} = \beta_{0\overline{SMK}}$ equals:

$$T_{28} = \frac{\hat{\beta}_{0\overline{SMK}} - \hat{\beta}_{0SMK}}{S(\hat{\beta}_{0\overline{S}} - \hat{\beta}_{0S})} = \frac{49.312 - 79.255}{\sqrt{583.528}} = -1.24 \text{, with } 0.20 < p < 0.30.$$

At $\alpha = 0.05$ we do not reject H_0 and conclude that the two intercepts are equal.

d From the above tests, we would conclude that the straight lines for smokers and nonsmokers are coincident since both tests failed to reject H_0.

5. a For NY: $\hat{Y} = 2.174 + 1.177X$

For CA: $\hat{Y} = 8.030 + 1.036X$

b $H_0: \beta_{1NY} = \beta_{1CA}$ $\quad H_a: \beta_{1NY} > \beta_{1CA}$

$$S^2_{p,Y|X} = \frac{(18*11.101) + (15*13.492)}{33} = 12.188$$

$$S^2_{(\hat{\beta}_{1NY} - \hat{\beta}_{1CA})} = 12.188\left[\frac{1}{(19)(76.870)} + \frac{1}{(16)(97.335)}\right] = 0.016$$

$$T_{33} = \frac{1.177 - (1.036)}{\sqrt{0.016}} = 1.115 \text{, with } 0.10 < p < 0.15.$$

At $\alpha = 0.05$, we do not reject H_0 and conclude that the slopes are the same for NY and CA.

c $H_0: \beta_{0NY} = \beta_{0CA}$ $\quad H_a: \beta_{0NY} > \beta_{0CA}$

$$S^2_{(\hat{\beta}_{0NY} - \hat{\beta}_{0CA})} = 12.188\left[\frac{1}{20} + \frac{1}{17} + \frac{(36.99)^2}{(19)(76.870)} + \frac{(36.371)^2}{(16)(97.335)}\right] = 23.09$$

$$T_{33} = \frac{2.174 - 8.030}{\sqrt{23.09}} = -1.219 \text{, with } 0.85 < p < 0.90.$$

At $\alpha = 0.05$, we do not reject H_0 that the two intercepts are equal for NY and CA.

d Since the tests for equal slopes and equal intercepts did not lead to rejection, we can conclude that the lines are coincident.

e $H_0: \rho_{NY} = \rho_{CA}$ $\quad H_a: \rho_{NY} \neq \rho_{CA}$

Fisher's $Z_{(NY)} = \dfrac{1}{2} \ln\dfrac{1 + (0.954)}{1 - (0.954)} = 1.874$

Fisher's $Z_{(CA)} = \dfrac{1}{2} \ln\dfrac{1 + (0.945)}{1 - (0.945)} = 1.783$

Then the test statistic equals:

$$Z = \frac{1.874 - (1.783)}{\sqrt{\frac{1}{17} + \frac{1}{14}}} = 0.252 \text{, with } p = 0.803.$$

At $\alpha = 0.05$, we do not reject H_0 and conclude that the correlation coefficients for each straight line regression are not significantly different.

9. **a** SBP = $\beta_0 + \beta_1$AGE + β_2QUET + β_3SMK + β_4AGE*SMK + β_5QUET*SMK + E
Smokers: SBP = $(\beta_0 + \beta_3) + (\beta_1 + \beta_4)$AGE + $(\beta_2 + \beta_5)$QUET + E
Nonsmokers: SBP = $\beta_0 + \beta_1$AGE + β_2QUET + E

b Smokers: $\hat{SBP} = 48.076 + 1.466(AGE) + 6.744(QUET)$

Nonsmokers: $\hat{SBP} = 48.613 + 1.029(AGE) + 10.451(QUET)$

c $H_0: \beta_4 = \beta_5 = 0$ H_a: at least one $\beta_i \neq 0$
Multiple Partial

$$F(QUET*SMK, AGE*SMK \mid AGE, QUET, SMK)_{2,26} = \frac{\frac{4915.630 - 4889.825}{2}}{58.09} = 0.222$$

with $p > 0.25$.
At $\alpha = 0.05$, we do not reject H_0 and conclude the two lines are coincident.

d $H_0: \beta_3 = \beta_4 = \beta_5 = 0$ H_a: at least one $\beta_i \neq 0$
Multiple Partial

$$F(SMK, QUET*SMK, AGE*SMK \mid AGE, QUET)_{3,26} = \frac{\frac{4915.630 - 4120.592}{3}}{58.09} = 4.562$$

with $0.01 < p < 0.025$.
At $\alpha = 0.05$, we reject H_0 and conclude the two lines are not coincident.

13. **a** R = 1 and TD = 1: $\hat{SBPL} = (\hat{\beta}_0 + \hat{\beta}_2 + \hat{\beta}_4 + \hat{\beta}_9) + \hat{\beta}_1(SBP1) + (\hat{\beta}_6 + \hat{\beta}_{11})RW$

R = 0 and TD = 1: $\hat{SBPL} = (\hat{\beta}_0 + \hat{\beta}_3 + \hat{\beta}_4 + \hat{\beta}_7) + \hat{\beta}_1(SBP1) + (\hat{\beta}_6 + \hat{\beta}_{13})RW$

R = -1 and TD = 1: $\hat{SBPL} = (\hat{\beta}_0 - \hat{\beta}_2 - \hat{\beta}_3 + \hat{\beta}_4 - \hat{\beta}_7 - \hat{\beta}_9) + \hat{\beta}_1 SBP1 + (\hat{\beta}_6 - \hat{\beta}_{11} - \hat{\beta}_{13})RW$

R = 1 and TD = 0: $\hat{SBPL} = (\hat{\beta}_0 + \hat{\beta}_2 + \hat{\beta}_5 + \hat{\beta}_{10}) + \hat{\beta}_1(SBP1) + (\hat{\beta}_6 + \hat{\beta}_{12})RW$

R = 0 and TD = 0: $\hat{SBPL} = (\hat{\beta}_0 + \hat{\beta}_3 + \hat{\beta}_5 + \hat{\beta}_8) + \hat{\beta}_1(SBP1) + (\hat{\beta}_6 + \hat{\beta}_{14})RW$

R = -1 and TD = 0: $\hat{SBPL} = (\hat{\beta}_0 - \hat{\beta}_2 - \hat{\beta}_3 + \hat{\beta}_5 - \hat{\beta}_8 - \hat{\beta}_{10}) + \hat{\beta}_1(SBP1) + (\hat{\beta}_6 - \hat{\beta}_{12} - \hat{\beta}_{14})RW$

R = 1 and TD = -1: $\hat{SBPL} = (\hat{\beta}_0 + \hat{\beta}_2 - \hat{\beta}_4 - \hat{\beta}_5 - \hat{\beta}_9 - \hat{\beta}_{10}) + \hat{\beta}_1(SBP1) + (\hat{\beta}_6 - \hat{\beta}_{11} - \hat{\beta}_{12})RW$

R = 0 and TD = -1: $\hat{SBPL} = (\hat{\beta}_0 + \hat{\beta}_3 - \hat{\beta}_4 - \hat{\beta}_5 - \hat{\beta}_7 - \hat{\beta}_8) + \hat{\beta}_1(SBP1) + (\hat{\beta}_6 - \hat{\beta}_{13} - \hat{\beta}_{14})RW$

R = -1 and TD = -1: $S\hat{B}PL = (\hat{\beta}_0 - \hat{\beta}_2 - \hat{\beta}_3 - \hat{\beta}_4 - \hat{\beta}_5 + \hat{\beta}_7 + \hat{\beta}_8 + \hat{\beta}_9 + \hat{\beta}_{10}) + \hat{\beta}_1(SBP1)$

$+ (\hat{\beta}_6 + \hat{\beta}_{11} + \hat{\beta}_{12} + \hat{\beta}_{13} + \hat{\beta}_{14})RW$

b $H_0: \beta_{11} = \beta_{12} = \beta_{13} = \beta_{14} = 0$ H_a: at least one $\beta_i \neq 0$.

$$F(X_{11}, X_{12}, X_{13}, X_{14}|X_1..., X_{10})_{4,89} = \frac{\frac{4.8892}{4}}{\frac{77.2303}{89}} = 1.409 \text{ , with } 0.1 < p < 0.25.$$

At $\alpha = 0.05$, we do not reject H_0 and conclude the lines are parallel.

c H_0: The three regression lines corresponding to rural, town, and urban background are parallel
i.e. ($H_0: \beta_7 = \beta_8 = \beta_9 = \beta_{10} = \beta_{11} = \beta_{12} = \beta_{13} = \beta_{14} = 0$).
H_a: at least one $\beta_i \neq 0$.

$$F(X_7,...,X_{14}| X_1...X_6)_{8,89} = \frac{\frac{7.6557}{8}}{\frac{77.2303}{89}} = 1.103 \text{ , with } p > 0.25.$$

At $\alpha = 0.05$, we do not reject H_0 and conclude that the three regression lines are parallel.

d $H_0: \beta_2 = \beta_3 = \beta_4 = \beta_5 = \beta_7 = \beta_8 = \beta_9 = \beta_{10} = \beta_{11} = \beta_{12} = \beta_{13} = \beta_{14} = 0$ H_a: at least one $\beta_i \neq 0$.

$$F(X_2, X_3, X_4, X_5, X_7,...,X_{14}| X_1, X_6)_{12,89} = \frac{\frac{SS\ Regression(full\ model) - SS\ Regression(X_1, X_6)}{12}}{MS\ Residual(full\ model)}$$

17. a $Y = \beta_0 + \beta_1 X + \beta_2 Z + \beta_3 XZ + E$

where $Z = 1$ if cool, 0 if warm.

b For cool: $\hat{Y} = 104.003 + 2.465X$

For warm: $\hat{Y} = 96.830 + 3.485X$

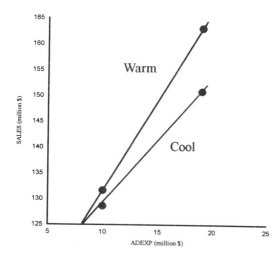

c $H_0: \beta_2 = \beta_3 = 0$ H_a: at least one $\beta_i \neq 0$.

$$F(Z, XZ| X)_{2,14} = \frac{\frac{294.130}{2}}{3.512} = 41.875 \text{ , with } p < 0.001.$$

At $\alpha = 0.05$, we reject H_0 and conclude the lines are not coincident.

d $H_0: \beta_3 = 0$ $H_a: \beta_3 \neq 0$.

$$F(XZ| X, Z)_{1,14} = \frac{34.063}{3.512} = 9.699 \text{ , with } p = 0.0076.$$

At $\alpha = 0.05$, we reject H_0 and conclude the lines are not parallel.

e Baseline sales are higher during the warm season relative to the cool season. Advertising expenditures are higher in the cool season relative to the warm season. By spending more money in advertising during the cool season, retailers are able to surpass the sales revenue of the warm season.

21. a $Y = \beta_0 + \beta_1 X + \beta_2 Z + \beta_3 XZ + E$

where $Z = 1$ if males; 0 if females.

b $H_0: \beta_2 = \beta_3 = 0$ $\qquad H_a:$ at least one $\beta_i \neq 0$.

$$F(Z, X_1Z | X_1)_{2,50} = \frac{\frac{19.652 - 13.098}{2}}{1.212} = 2.704 \text{ , with } 0.05 < p < 0.10.$$

At $\alpha = 0.05$, we do not reject H_0 and conclude the lines are coincident.

c $H_0: \beta_3 = 0$ $\qquad H_a: \beta_3 \neq 0$.

$$F(X_1Z | X_1, Z)_{1,50} = \frac{4.348}{1.212} = 3.587 \text{ , with } p = 0.064.$$

At $\alpha = 0.05$, we do not reject H_0 and conclude the lines are parallel.

d The change in refraction-baseline refractive relationship is the same for males and females.

Chapter 15

1. a $SBP = \beta_0 + \beta_1(QUET) + \beta_2 SMK + E$

b For smokers: $SBP = (\beta_0 + \beta_2) + \beta_1(QUET) + E$

$\overline{SBP}(adj) = (63.876 + 8.571) + (22.116)(3.441) = 148.548$

For nonsmokers: $SBP = \beta_0 + \beta_1(QUET) + E$

$\overline{SBP}(adj) = 63.876 + (22.116)(3.441) = 139.977$

c $H_0: \beta_2 = 0 \qquad H_a: \beta_2 \neq 0$ in the model $SBP = \beta_0 + \beta_1(QUET) + \beta_2 SMK + E$

$T_{29} = 2.707, \quad p = 0.011$.
At $\alpha = 0.05$ we reject H_0 conclude that mean SBP is different for smokers and nonsmokers after adjusting for QUET.

d Finding the 95% confidence interval for the true difference in adjusted mean SBP is equivalent to finding the 95% confidence interval for $\hat{\beta}_2$.

i.e. $\overline{SBP}_{SMK}(adj) - \overline{SBP}_{NON}(adj) = 8.571 = \hat{\beta}_2$

The 95% confidence interval for $\hat{\beta}_2$ equals:

$\hat{\beta}_2 \pm t_{29,\, 0.975} * S_{\hat{\beta}_2} = 8.571 \pm 2.045 * 3.167 = (2.094, 15.048)$.

5. a $VIAD = \beta_0 + \beta_1 IQM + \beta_2 IQF + \beta_3 Z + E$

where $Z_1 = 1$ if female, 0 male.

b For males: VIAD (adj) = -3.307 vs. -3.00 unadjusted.
For females: VIAD (adj) = 1.889 vs. 1.60 unadjusted.

c $H_0: \beta_3 = 0 \qquad H_a: \beta_3 \neq 0$ in the model $VIAD = \beta_0 + \beta_1 IQM + \beta_2 IQF + \beta_3 Z + E$

$T_{16} = 1.659, \quad p = 0.117$.
At $\alpha = 0.05$, we do not reject H_0 and conclude the mean scores do not significantly differ by gender after adjusting for IQM and IQF.

d 95% CI: $\hat{\beta}_3 \pm t_{16,\, 0.975} * S_{\hat{\beta}_3} = 5.196 \pm 2.120 * 3.133 = (-1.446, 11.838)$

9. **a** $LN\_BRNTL = \beta_0 + \beta_1 WGT + \beta_2 Z_1 + \beta_3 Z_2 + \beta_4 Z_3 + E$

   where $Z_1 = 1$ if 100 ppm, 0 otherwise
   $Z_2 = 1$ if 500 ppm, 0 otherwise
   $Z_3 = 1$ if 1000 ppm, 0 otherwise.

   b $\hat{\beta}_0 = -0.764 \quad \hat{\beta}_1 = 0.0006 \quad \hat{\beta}_2 = 0.828$

   $\hat{\beta}_3 = 3.571 \quad \hat{\beta}_4 = 4.214$

   c
PPM_TOLU	Adjusted Means	Unadjusted Means
50	-0.537	-0.548
100	0.291	0.282
500	3.034	3.019
1000	3.677	3.668

   d $H_0 : \beta_2 = \beta_3 = \beta_4 = 0$
   H_a: at least one $\beta_i \neq 0$ (i=2,3,4) in the model
   $LN\_BRNTL = \beta_0 + \beta_1 WGT + \beta_2 Z_1 + \beta_3 Z_2 + \beta_4 Z_3 + E$

   $$F(Z_1, Z_2, Z_3 | WGT)_{3,55} = \frac{\frac{189.528}{3}}{0.038} = 1662.526 \text{, with } p < 0.001.$$

   At $\alpha = 0.05$, we reject H_0 and conclude that the adjusted means are significantly different.

13. **a** $Y = \beta_0 + \beta_1 AGE + \beta_2 Z + E$
    b $Y = \beta_0 + \beta_1 AGE + \beta_2 Z + \beta_3 AGE*Z + E$
    $H_0 : \beta_3 = 0 \qquad H_a : \beta_3 \neq 0$ (Full model: $Y = \beta_0 + \beta_1 AGE + \beta_2 Z + \beta_3 AGE*Z + E$)
    $T_{26} = -1.677$, p = 0.106.
    At $\alpha = 0.05$, we do not reject H_0 and conclude the ANACOVA model in part (a) is appropriate.

    c
Location	Adjusted Means	Unadjusted Means
Intown/inner	81.526	82.187
Outer	85.46	84.807

    $H_0 : \beta_2 = 0 \qquad H_a : \beta_2 \neq 0$
    $T_{27} = 1.212$, p = 0.236.
    At $\alpha = 0.05$, we do not reject H_0 and conclude the adjusted means do not significantly differ.

Chapter 16

1. a <u>Forward selection method:</u> The dependent variable is Y = WGT and the five independent variables are
X_1 = HGT, X_2 = AGE, X_3 = (AGE)^2, X_4 = (HGT)^2, X_5 = (AGE*HGT)
For the forward selection method, we execute the following steps:
<u>Step 1:</u> We find the simple correlation coefficients between the individual X_i and Y, and then select the variable which has the highest correlation coefficient to enter the model first (equivalently, we could examine the R^2 values in the output provided). Here we find that:
r_{YX1}= 0.814, r_{YX2}= 0.770, r_{YX3}= 0.767, r_{YX4}= 0.817, r_{YX5}= 0.868
The highest correlation here is r_{YX5}= 0.868. Thus the first variable to enter the model is X_5. Our current model is:

$$\hat{WGT} = 37.600 + 0.053(AGE*HGT)$$

We then calculate $F(X_5)$:

$$(X_5)_{1,10} = \frac{SSR(X_5)}{MSE(X_5)} = \frac{SSY-(10)MSE(X_5)}{MSE(X_5)} = \frac{888.25 - (10)(21.8918)}{21.892} = \frac{669.332}{21.892} = 30.5$$

with p < 0.001.
The F statistic is significant so the variable is included and proceed to step 2.
<u>Step 2:</u> We compute the partial F statistics associated with each of the remaining variables using the model formed in Step 1.

$F(X_4| X_5)_{1,9} = 0.69$ $F(X_1| X_5)_{1,9} = 0.88$

$F(X_2| X_5)_{1,9} = 0.51$ $F(X_3| X_5)_{1,9} = 0.78$

The largest partial F statistic is $F(X_1| X_5) = 0.88$ which is not significant at $\alpha = 0.10$, so we stop and do not add any more variables to the model. The final model by forward selection is:

$$\hat{WGT} = 37.600 + 0.053(AGE*HGT) \quad \text{with } R^2 = 0.754.$$

b <u>Backward Elimination Method:</u>
<u>Step 1:</u> We determine the fitted regression equation containing all of the independent variables. We find:

$$\hat{GT}= -91.937 + 5.297(HGT) - 1.797(AGE) - 0.011(AGE^2) - 0.054(HGT^2) + 0.093(AGE*HGT)$$

with the ANOVA table:

Source	df	SS	MS
Regression	5	714.383	142.877
Residual	6	173.867	28.978
Total	11	888.250	

Step 2: We compute partial F statistics for each variable in the model as though it were the last variable to enter
Partial F (based on 1 and 6 df)
$F(X_1| X_2, X_3, X_4, X_5) = 0.834$
$F(X_2| X_1, X_3, X_4, X_5) = 0.034$
$F(X_3| X_1, X_2, X_4, X_5) = 0.000$
$F(X_4| X_1, X_2, X_3, X_5) = 0.736$
$F(X_5| X_1, X_2, X_3, X_4) = 0.113$
Step 3: The lowest partial F is for X_3. Since it is not significant at $\alpha = 0.10$, we remove X_3 from the model and recompute the fitted regression using $X_1, X_2, X_4,$ and X_5. The model obtained is:

$$\hat{WGT} = -92.433 + 5.323(HGT) - 1.849(AGE) - 0.054(HGT^2) + 0.091(AGE*HGT)$$

Step 4: We compute partial F statistics for the remaining variables.
Partial F (based on 1 and 7 df)
$F(X_1| X_2, X_4, X_5) = 1.14$
$F(X_2| X_1, X_4, X_5) = 0.05$
$F(X_4| X_1, X_2, X_5) = 0.86$
$F(X_5| X_1, X_2, X_4) = 0.30$
Step 5: The smallest F is for X_2, which is not significant at $\alpha = 0.10$, we remove X_2 from the model and recompute the fitted regression using $X_1, X_4,$ and X_5. The model obtained is:

$$\hat{WGT} = -89.236 + 4.861(HGT) - 0.046(HGT^2) + 0.053(AGE*HGT)$$

Step 6: We compute partial F statistics for the remaining variables.
Partial F (based on 1 and 8 df)
$F(X_1| X_4, X_5) = 1.28$
$F(X_4| X_1, X_5) = 1.11$
$F(X_5| X_1, X_4) = 5.46$
Step 7: The smallest F is for X_4, which is not significant at $\alpha = 0.10$, we remove X_4 from the model and recompute the fitted regression using X_1 and X_5. The model obtained is:

$$\hat{WGT} = 25.062 + 0.368(HGT) + 0.039(AGE*HGT)$$

Step 8: We compute partial F statistics for the remaining variables.
Partial F (based on 1 and 9 df)
$F(X_1| X_5) = 0.88$
$F(X_5| X_1) = 4.51$
Step 9: The smallest F is for X_1, which is not significant at $\alpha = 0.10$, we remove X_4 from the model and recompute the fitted regression using X_1 and X_5. The model obtained is:

$$\hat{WGT} = 37.600 + 0.053(AGE*HGT)$$

$F(X_5)_{1,11} = 30.57$ with $p < 0.001$ so we stop and our final model is the same model obtained for (a).

    c   Starting with HGT and AGE alone, we find that HGT is a more important predictor than AGE, but that the addition of AGE to a model already containing HGT is significant. Thus, both HGT and AGE would be included in the model at this stage. We would further find that no second order term among $(AGE)^2$, $(HGT)^2$, and AGE*HGT significantly adds to predicting WGT once HGT and AGE are in the model. The best model resulting from this approach is:

$$\hat{WGT} = 6.553 + 0.722(HGT) + 2.050(AGE) \text{ with } R^2 = 0.780.$$

    d   The first two approaches result in the model including only the AGE*HGT interaction. It is difficult to interpret results for such a model, since the variables which make up the interaction term are not included. The third approach results in a model which is more easily interpreted. It is possible to conduct modified forward and backwards stepwise strategies in which interaction terms are only considered for inclusion if the terms that make up the interaction term are already in the model.

3. Stepwise Regression Method
    i   For smokers:
Using the SAS output, the final estimated model is:

$$\hat{SBP} = 102.220 + 0.252(AGE*QUET)$$

    ii   For non-smokers:
The final estimated model is:

$$\hat{SBP} = 93.073 + 0.250(AGE*QUET)$$

These two models are different from those obtained in 2 (d) by putting SMK = 1 (for smokers) and SMK = 0 (for non-smokers). Doing so for the equations in 2 (d) we get the models:

For smokers: $\hat{SBP} = 88.915 + 0.482(AGE) + 0.186(AGE*QUET)$

For nonsmokers: $\hat{SBP} = 79.096 + 0.482(AGE) + 0.186(AGE*QUET)$

5. The best regression models using the sequential procedure of adding AGE first for Females and Males are:

Females: $\hat{DEP} = 190.012 - 1.099AGE - 1.217MC$

Males: $\hat{DEP} = 270.056 - 2.514AGE - 1.065MC$

For females, the above model was selected as best because of its high r^2 (0.401), satisfactory C(p) (2.238) and low MSE (3274.364). The next best model was the full model with $r^2 = 0.413$, C(p) = 4.0, MSE = 3478.318. Emphasizing parsimony we find that the above model is more favorable than the full model.

For males, the above model was selected as best because of its high r^2 (0.321), satisfactory C(p) (2.0) and low MSE (1957.856). The next best model was the full model with $r^2 = 0.322$, C(p) = 4.0, MSE = 2060.875. Emphasizing parsimony, we find that the above model is more favorable than the full model.

7. a Using the C(p) criterion exclusively, one finds that the three models with the most favorable C(p) values contain AGE alone (1.23), AGE and WGT (2.63), or all three variables (4). None of these models are particularly impressive, and since it would be hard to argue that either of the multiple variable models is better than the model with AGE alone, one could take AGE alone as the best model. Upon further investigation, the difficulty is seen to be partially due to none of these models having a significant overall F test.
    b Since there seems to be no rational for grouping the variables (age, weight, and height) in any way, then chunks are taken to be the three variable specific pairs of linear/quadratic terms (i.e. AGE_C and AGE_CSQ; HGT_C and HGT_CSQ; WGT_C and WGT_CSQ, where the _C terms are the centered variables, and the _CSQ terms are the squared centered terms). A plausible forward chunkwise strategy is to treat each chunk/pair as a distinct entity that cannot be split and proceed in the usual forward manner (use $\alpha = 0.10$):
    Step 1: WGT_C, WGT_CSQ added to the model.   F test = 2.78
    Step 2: AGE_C, AGE_CSQ added to the model. F test = 3.49
    Step 3: Stop HGT_C, HGT_CSQ not significant.   F test = 0.16
    c Again focusing on C(p), the all possible regressions method yields the following 'best' models for each of the model sizes:

Number in Model	Variables	C(p)	R^2	MSE
1	WGT_CSQ	8.026	0.078	0.771
2	WGT_C, WGT_CSQ	5.213	0.300	0.631
3	WGT_C, WGT_CSQ, AGE_CSQ	4.337	0.432	0.554
4	WGT_C, WGT_CSQ, AGE_C, AGE_CSQ	3.312	0.571	0.456
5	WGT_C, WGT_CSQ, AGE_C, AGE_CSQ, HGT_CSQ	5.225	0.576	0.497
6	Full Model	7.00	0.586	0.539

The best model is most likely the four variable model including WGT_C, WGT_CSQ, AGE_C, and AGE_CSQ. The R^2, C(p) and MSE for this model are better than for any of the smaller models; and they are similar to, if not better than, the statistics for the larger models, which of course are less parsimonious.
    d Any model containing only first order terms (part (a)) is seriously deficient. The model in parts (b) and (c) is the best one.

9. a A model containing X_1, X_3, and X_4 is the best model by this method. The model has a relatively high R^2, a satisfactory C(p), and one of the lowest MSEs. The model also has the benefit of parsimony compared to the larger models.
    b The selected model contains X_1, X_3, and X_4.
    c The selected model contains X_1, X_3, and X_4.
    d All three methods selected the same model and it appears to be the best model for reasons cited in (a).

11. a The model containing X_1 and X_3, would be the recommended model. Its R^2, C(p) and MSE are clearly are superior to the best one-variable model statistics, and they are similar to the statistics for the full model. The full model, however, is not parsimonious.
    b The results of the stepwise regression show that the model containing X_1 and X_3 is the best model.
    c $Y = \beta_0 + \beta_1 X_1 + \beta_3 X_3 + E$ appears to be best, given the results of parts (a) and (b).

Chapter 17

1. a

Treatment	Mean	S(Y)
1	7.5	1.643
2	5	1.265
3	4.333	1.033
4	5.167	1.472
5	6.167	2.041

$\overline{Y} = 5.633$

b ANOVA table

Source	df	SS	MS	F
Treatment	4	36.467	9.117	3.90
Error	25	58.50	2.34	
Total	29	94.967		

c $H_0: \mu_1 = \mu_2 = \mu_3 = \mu_4 = \mu_5 = 0$ H_a: at least two treatments have different population means.

$$F_{4,25} = \frac{9.117}{2.34} = 3.896, \text{ with } p = 0.0136.$$

At $\alpha = 0.05$ we reject H_0 and conclude that at least two treatments have different population means

d Estimates of true effects $(\mu_i - \mu)$ where μ is the overall mean, i.e.

$$\mu = \frac{1}{5}\sum_{i=1}^{5} \mu_i, \; \mu_i = \text{the population mean for the } i^{th} R_x:$$

Treatment i	$(\overline{Y}_i - \overline{Y})$
1	1.8667
2	-0.6333
3	-1.3000
4	-0.4667
5	0.5333

Total $\sum_{i=1}^{5} (\overline{Y}_i - \overline{Y})$ 0.000

e We define X_i such that
$X_i = 1$, for R_x i ; 0, otherwise
where i = 1, 2, 3, 4 . Then the appropriate regression model is
$Y = \beta_0 + \alpha_1 X_1 + \alpha_2 X_2 + \alpha_3 X_3 + \alpha_4 X_4 + E$
where the regression coefficients are as follows:
$\beta_0 = \mu_5$,
$\alpha_1 = \mu_2 - \mu_5, \alpha_2 = \mu_2 - \mu_5, \alpha_3 = \mu_3 - \mu_5, \alpha_4 = \mu_4 - \mu_5$
For $X_i = -1$, for R_x 5 ; 1, for R_x i ; 0, otherwise
the regression coefficients are:
$\beta_0 = \mu, \alpha_1 = \mu_1 - \mu, \alpha_2 = \mu_2 - \mu, \alpha_3 = \mu_3 - \mu, \alpha_4 = \mu_4 - \mu$, and also
$-(\alpha_1 + \alpha_2 + \alpha_3 + \alpha_4) = \mu_5 - \mu$

f Calculating by hand: We rank the sample means in descending order of magnitude and get
$$\bar{Y}_1 > \bar{Y}_5 > \bar{Y}_4 > \bar{Y}_2 > \bar{Y}_3$$
So the order of comparisons to be made is 1 vs 3, 1 vs 2, 1 vs 4, 1 vs 5, 5 vs 3, 5 vs 2, 5 vs 4, 4 vs 3, 4 vs 2, 2 vs 3.

Now MSE= 2.34 (from the ANOVA given in (b)), and

$$n_1 = n_2 = n_3 = n_4 = n_5 = 6 = n^*, \; n = \sum_{i=1}^{5} n_i = 30$$

$k = 5, \; \alpha = 0.05$

Scheffè method, $S^2 = (k-1) F_{k-1, n-k; 1-\alpha}$
$= 4 * F_{4, 25; 0.95}$
$= 4 * 2.75$
$= 11$
$S = 3.317$

Thus the half width, w_s, by Scheffè method is

$$w_s = S \sqrt{MSE \left(\frac{1}{6} + \frac{1}{6}\right)}$$

$$= 3.317 \sqrt{2.34 \left(\frac{1}{3}\right)}$$

$$= 2.929$$

For Tukey's method $\quad T = \dfrac{1}{\sqrt{n^*}} q_{k, \; n-k; \; 1-\alpha}$

$$= \frac{1}{\sqrt{6}} * q_{5, \; 25, \; 0.95}$$

$$= \frac{1}{\sqrt{6}} * 4.158$$

$$= 1.697$$

Thus the half width, w_T, by Tukey's method is:

$$w_T = T\sqrt{MSE}$$

$$= 1.697 * \sqrt{2.34}$$

$$= 2.596$$

For LSD method, total number of pairwise combinations = $C_2^5 = 10$,

so that $\alpha^* = \dfrac{\alpha^*}{10} = \dfrac{0.05}{10} = 0.005$; and $\dfrac{\alpha^*}{2} = 0.0025$. Hence the half width

W, by LSD method is

$$W_L = (t_{25, 1 - 0.0025}) \sqrt{MSE\left(\frac{1}{6} + \frac{1}{6}\right)}$$

$$= 2.827$$

Confidence intervals: Pairwise differences between the sample means for these treatments are as follows:

$\overline{Y_1} - \overline{Y_3} = 3.167$, $\overline{Y_1} - \overline{Y_2} = 2.5$, $\overline{Y_1} - \overline{Y_4} = 2.333$, $\overline{Y_1} - \overline{Y_5} = 1.333$, $\overline{Y_5} - \overline{Y_3} = 1.834$,

$\overline{Y_5} - \overline{Y_2} = 1.167$, $\overline{Y_5} - \overline{Y_4} = 1.000$, $\overline{Y_4} - \overline{Y_3} = 0.834$, $\overline{Y_4} - \overline{Y_2} = 0.167$, and $\overline{Y_2} - \overline{Y_3} = 0.667$.

Confidence intervals for the two largest differences are given as follows:

Comparison	Scheffe'	Tukey	LSD
$(\mu_1 - \mu_3)$	$(\overline{Y_1} - \overline{Y_3}) \pm W_S$	$(\overline{Y_1} - \overline{Y_3}) \pm W_T$	$(\overline{Y_1} - \overline{Y_3}) \pm W_L$
	$= 3.167 \pm 2.929$	$= 3.167 \pm 2.597$	$= 3.167 \pm 2.827$
	$= (0.238, 6.096)$	$= (0.57, 5.764)$	$= (0.34, 5.994)$

Conclusion: Treatment 1 and 3 are significantly different using an overall $\alpha = 0.05$.

$(\mu_1 - \mu_2)$	$(\overline{Y_1} - \overline{Y_2}) \pm W_S$	$(\overline{Y_1} - \overline{Y_2}) \pm W_T$	$(\overline{Y_1} - \overline{Y_2}) \pm W_L$
	$= 2.5 \pm 2.929$	$= 2.5 \pm 2.597$	$= 2.5 \pm 2.827$
	$= (-0.429, 5.429)$	$= (-0.097, 5.097)$	$= (-0.327, 5.327)$

Conclusion: Since 0 is included in the interval, treatments 1 and 2 are not significantly different. Also, we conclude that all other remaining comparisons, which involve smaller (in absolute value) pairwise mean differences, are not significant using $\alpha = 0.05$. Thus we conclude that with an overall $\alpha = 0.05$, treatments 1 and 3 are significantly different, but that all other pairwise comparisons are non-significant. (Note that the SAS output can be used, instead of hand calculation, to arrive directly at this conclusion without the intermediate calculations).

Of the three methods, Scheffe's method gives the widest interval, the LSD being the next widest, and Tukey's method gives the narrowest interval.

5. a

$$SS\ Parties = \frac{(85*5)^2 + (80*5)^2 + (95*5)^2 + (50*5)^2}{5} -$$

$$\frac{[(85*5) + (80*5) + (95*5) + (50*5)]^2}{20} = 125750 - 120125 = 5625$$

$$MSE = \frac{(n_1 - 1)S_1^2 + (n_2 - 1)S_2^2 + (n_3 - 1)S_3^2 + (n_4 - 1)S_4^2}{(n_1 + n_2 + n_3 + n_4 - 4)} = 50.25$$

Also, SSE = 50.25 * 16 = 804, noting that the df associated with error = total df - 3 = 19 - 3 = 16. Thus we have the ANOVA tables as:

Source	df	SS	MS	F
Parties	3	5625	1875.00	37.31
Error	16	804	50.25	
Total	19	6429		

b $H_0: \mu_1 = \mu_2 = \mu_3 = \mu_4$ H_a: at least two parties have different mean authoritarianism scores.
$F_{(3,16)} = 37.31$, with $p < 0.001$.
At $\alpha = 0.05$, we reject H_0 and conclude that there are significant differences in the authoritarianism scores of the members of different political parties.

c Regression model:
$$Y = \beta_0 + \alpha_1 X_1 + \alpha_2 X_2 + \alpha_3 X_3 + E$$
where $X_i = $ 1 if party # i, 0 otherwise; i = 1, 2, 3

d Tukey's method: We arrange the party means in descending order of magnitude,
$$\overline{Y}_3 > \overline{Y}_1 > \overline{Y}_2 > \overline{Y}_4$$
so we shall compare parties in the following sequence:
3 vs 4, 3 vs 2, 3 vs 1, 1 vs 4, 1 vs 2, 2 vs 4.
Here $n_1 = n_2 = n_3 = n_4 = n^* = 5$, $k = 4$, $\alpha = 0.05$, $n = \Sigma n_i = 20$. Also,

$$T = \frac{1}{\sqrt{n*}} * q_{K, n-k, 1-\alpha}$$

$$= \frac{1}{\sqrt{5}} * 4.08$$

$$= 1.8246$$

Hence the half-width of the confidence interval by Tukey's method is
$$W_T = T * \sqrt{MSE}$$
$$= 1.8246 * \sqrt{50.25}$$
$$= 12.93$$

The confidence intervals for different comparisons are then given as follows:

Comparison	Confidence Interval	Remark
P_3 vs P_4:	$(\overline{Y}_3 - \overline{Y}_4) \pm 12.93 = (32.07, 57.93)$	Significant
P_3 vs P_2:	$(\overline{Y}_3 - \overline{Y}_2) \pm 12.93 = (2.07, 27.93)$	Significant
P_3 vs P_1:	$(\overline{Y}_3 - \overline{Y}_1) \pm 12.93 = (-2.93, 22.93)$	Not Significant since 0 in interval
P_1 vs P_4:	$(\overline{Y}_1 - \overline{Y}_4) \pm 12.93 = (22.07, 47.93)$	Significant
P_1 vs P_2:	$(\overline{Y}_1 - \overline{Y}_2) \pm 12.93 = (-7.93, 17.93)$	Not Significant
P_2 vs P_4:	$(\overline{Y}_2 - \overline{Y}_4) \pm 12.93 = (17.07, 42.93)$	Significant

Thus at $\alpha = 0.05$, all differences except for $(\overline{Y}_1 - \overline{Y}_2)$ and $(\overline{Y}_3 - \overline{Y}_1)$ are significant.

9. a $H_0: \mu_1 = \mu_2 = \mu_3$ H_a: at least two mean generation times differ.
$F_{(3,20)} = 46.99$, with $p < 0.001$.
At $\alpha = 0.05$, we reject H_0 and conclude that there are significant differences between strains.

b

Source	df	SS	MS	F
Strains	3	134713.00	44904.33	46.99
Error	20	19113.244	955.662	
Total	23	153826.244		

c Tukey's method of multiple comparison:
We have
$$n^* = 6, k = 4, n = 24, \alpha = 0.05$$
so that
$$T = \frac{1}{\sqrt{n^*}} q_{k,n-k,1-\alpha} = \frac{1}{\sqrt{6}} * 3.96 = 1.6167$$

and the half-width, $W_T = T\sqrt{MSE} = 1.6167 * \sqrt{955.662} = 49.977$

Also, $\bar{Y}_C = 540.4 > \bar{Y}_D = 450.8 > \bar{Y}_A = 420.3 > \bar{Y}_B = 330.7$.
The confidence intervals for pairwise comparisons are:

Comparison	Confidence Interval	Remark
C vs B:	$(\bar{Y}_C - \bar{Y}_B) \pm 49.98 = (159.72, 259.68)$	Significant
C vs A:	$(\bar{Y}_C - \bar{Y}_A) \pm 49.98 = (70.12, 170.08)$	Significant
C vs D:	$(\bar{Y}_C - \bar{Y}_D) \pm 49.98 = (39.62, 139.58)$	Significant
D vs B	$(\bar{Y}_D - \bar{Y}_B) \pm 49.98 = (70.12, 170.08)$	Significant
D vs A	$(\bar{Y}_D - \bar{Y}_A) \pm 49.98 = (-19.48, 80.48)$	Not significant
A vs B	$(\bar{Y}_A - \bar{Y}_B) \pm 49.98 = (39.62, 139.58)$	Significant

Thus we conclude that
$$\mu_C > (\mu_D = \mu_A) > \mu_B$$
where $\mu_A, \mu_B, \mu_C,$ and μ_D are population means from Strains A, B, C, and D respectively.

13. a

Source	df	SS	MS	F
Dosage	3	566.628	188.876	14.71
w/ in Dosage	44	564.96	12.84	
Total	47	1131.588		

d = 3   since there are 4 dose levels
f = 47 since there are 48 subjects
e = 44 by subtraction (47 - 3 = 44).
g =   F · MS (w/ in dosage) = 14.71 * 12.84 = 188.876
a =   MS (dosage) · df (dosage) = 188.876 * 3 = 566.628
b =   MS (w/ in dosage) · df (w/ in dosage) = 12.84 * 44 = 564.96
c =   SS (dosage) + SS (w/ in dosage) = 1131.588

17. a $Y_{ij} = \mu + \alpha_i + E_{ij}$ i=1,2,3; j = 1,...,6.
    b MSM = 4.667   F = 5.53
       MSE = 0.844

- **c** $H_0: \mu_1 = \mu_2 = \mu_3$ H_a: There is a significant difference in attitude toward advertising by practice type.
  $F_{2,15} = 5.53$, with $p = 0.0159$.
  At $\alpha = 0.05$, we reject H_0 and conclude that there are significant differences in attitude toward advertising by practice type.
- **d** GP vs IM: $3.6667 - 2.00 \pm 1.4398$ (0.2269, 3.1065) Significant
  GP vs FP: $3.6667 - 2.333 \pm 1.4398$ (-0.1061, 2.7735) N.S.
  FP vs IM: $2.333 - 2.00 \pm 1.4398$ (-1.1068, 1.7728) N.S.
- **e** same as 17(a)
- **f**

Source	df	SS	MS	F
Model	2	26.333	13.167	20.43
Error	15	9.667	0.6444	
Total	17	35.999		

- **g** $H_0: \mu_1 = \mu_2 = \mu_3$ H_a: at least two means differ by practice type.
  $F_{2,15} = 20.43$, with $p = 0.0001$.
  At $\alpha = 0.05$, we reject H_0 and conclude that there is a significant difference in influence on prescription writing habits by practice type.
- **h** GP vs IM: $4.3333 - 1.5000 \pm 1.2578$ (1.5755, 4.0911) Significant
  GP vs FP: $4.3333 - 2.1667 \pm 1.2578$ (0.9088, 3.4244) Significant
  FP vs IM: $2.1667 - 1.500 \pm 1.2578$ (-0.5911, 1.9245) N.S.

21.
- **a** $Y_{ij} = \mu + \alpha_i + E_{ij}$ $i=1,..3;$ $j=1,...n_i$
  Clear zone sizes are fixed effect factors.
- **b**

Source	df	SS	MS	F	p-value
Model	2	14.704	7.352	5.462	0.0073
Error	48	64.592	1.346		
Total	50	79.296			

- **c** $H_0: \mu_1 = \mu_2 = \mu_3$ H_a: at least two means five-year changes differ by clear zone.
  $F_{2,48} = 5.462$, with $p = 0.0073$.
  At $\alpha = 0.05$, we reject H_0 and conclude that mean five-year changes significantly differ by clear zone size.
- **d** μ_1 vs μ_3: (0.3368, 2.2188) significant
  μ_1 vs μ_2: (-0.2562, 1.6603) not significant
  μ_2 vs μ_3: (-1.6603, 0.2562) not significant
  The mean five-year refractive changes significantly differ between 3.0 mm and 4.0 mm.

Chapter 18

1. **a** The rats are the blocks and the three chemicals are the treatments.
    b SSE = 25.0   MSE = 1.786
    Type I SS (chem) = 25.0   MS (chem) = 12.5
    c $H_0: \mu_1 = \mu_2 = \mu_3$ H_a: The mean irritation scores differ by chemical type.
    $F_{2, 14} = 7.00$, with $p = 0.0078$.
    At $\alpha = 0.05$ we reject H_0 and conclude that are significant differences in the toxic effects of the three chemicals.

    d   98% CI: $(6.25 - 7.5) \pm 2.624\sqrt{1.786(\frac{1}{8} + \frac{1}{8})} = (-3.0, 0.5)$

    e $R^2 = \dfrac{(SS \ chemical) + (SS \ rats)}{Total \ SS} = 0.635$

    f   Fixed effects ANOVA model:
    $$Y_{ij} = \mu + \tau_i + \beta_j + E_{ij} \ ; i = 1, 2, 3; j = 1, 2, ..., 8$$
    where Y_{ij} = observation on the jth rat for the ith chemical effect;
    μ = overall mean;
    τ_i = i^{th} chemical effect;
    β_j = j^{th} rat effect;
    E_{ij} = error due to observation on the jth rat for the ith chemical effect;
    and E_{ij}'s are independent and assumed to be normally distributed.
    Regression Model:

    $$Y = \beta_0 + \alpha_1 X_1 + \alpha_2 X_2 + \sum_{j=1}^{7} \beta_j Z_j + E$$

    where $X_1 = \begin{array}{l} 1 \text{ if chemical I} \\ 0 \text{ if chemical II} \\ -1 \text{ if chemical III} \end{array}$, $X_2 = \begin{array}{l} 0 \text{ if chemical I} \\ 1 \text{ if chemical II} \\ -1 \text{ if chemical III} \end{array}$, $Z_j = \begin{array}{l} -1 \text{ if rat 8} \\ 1 \text{ if rat } j \ (j = 1, 2..., 7) \\ 0 \text{ otherwise} \end{array}$

    g Assumptions: The assumptions underlying the model on which the validity of the analysis depends are;
    (i) additivity of the model (no interaction),
    (ii) homogeneity of variance,
    (iii) normality of the errors,
    (iv) independence of the errors.

5. $H_0: \mu_1 = \mu_2 = \mu_3$ H_a: At least two mean ESP scores differ by person.
    $F_{2, 8} = 15.12$, with $p = 0.0019$.
    At $\alpha = 0.05$ we reject H_0 and conclude that there are significant differences in ESP ability by person. Since the F test for blocking on days is not significant, one may conclude that blocking on days is not necessary.

9.

Source	df	SS	MS	F
Treatments	4	160	40	5.00
Blocks	5	240	48	6.00
Error	20	160	8	
Total	29	560		

    a   df(Blocks) = df(Error)/df(Treatments) = 20/4 = 5
    f   MS(Error) = MS(Blocks)/F(Blocks) = 48/6 = 8
    e   MS(Treatments) = F(Treatments)*df(Treatments) = 5*8 = 40
    b   SS(Treatments) = MS(Treatments)*df(Treatments) = 40*4 = 160
    c   SS(Blocks) = MS(Blocks)*df(Blocks) = 48*5 = 240
    d   SS(Error) = MS(Error)*df(Error) = 8*20 = 160
    g   Test of Treatments:
          $F_{4, 20} = 5.00$, with $0.005 < p < 0.01$.
          At $\alpha = 0.05$ we reject H_0 and conclude that there is a significant main effect of treatments.
          Test of Blocks:
          $F_{5, 20} = 6.00$, with $0.001 < p < 0.005$
          At $\alpha = 0.05$ we reject H_0 and conclude that there is a significant main effect of blocks.

Chapter 19

1. a The two factors are levorphanol and epinephrine.
   b Both factors should be considered fixed.
   c Rearrangement of the data into a 2-way ANOVA layout:

		Levels of Epinephrine	
		Absence −	Presence +
Levels of Levorphanol	Absence −	(Control) 1.90, 1.80, 1.54, 4.10, 1.89	(Epinephrine only) 5.33, 4.85, 5.26, 4.92, 6.07
	Presence +	(Levporphanol only) 0.82, 3.36, 1.64, 1.74, 1.21	(Levorphanol and Epinephrine) 3.08, 1.42, 4.54, 1.25, 2.57

   d Table of sample means:

		Epinephrine		Total (Row Mean)
		−	+	
Levorphanol	−	2.25	5.28	3.77
	+	1.75	2.57	2.16
Total (Col. Mean)		2.00	3.93	2.96

The presence of levorphanol appears to reduce stress, whereas the presence of epinephrine appears to increase stress. In the presence of epinephrine, levorphanol reduces stress by an average of 2.71 units, whereas in the absence of epinephrine, levorphanol reduces stress by only 0.50 units, suggesting the possibility of an interaction.

   e ANOVA table:

Source	df	SS	MS	F
Levorphenol	1	12.832	12.932	12.60
Epinephrine	1	18.586	18.586	18.25
Interaction	1	6.161	6.161	6.05
Error	16	16.298	1.019	
Total	19	53.877		

   f <u>Main effect of levorphenol</u>
   H_0: There is no significant main effect of levorphenol on stress.
   H_a: There is a significant main effect of levorphenol on stress.
   $F_{1,16} = 12.60$, $p = 0.0027$.
   At $\alpha = 0.05$ we reject H_0 and conclude that there is a significant main effect of levorphenol on stress.

   <u>Main effect of epinephrine</u>
   H_0: There is no significant main effect of epinephrine on stress.
   H_a: There is a significant main effect of epinephrine on stress.
   $F_{1,16} = 18.25$, $p = 0.0066$.
   At $\alpha = 0.05$ we reject H_0 and conclude that there is a significant main effect of epinephrine on stress.

Interaction
H_0: There is no significant interaction between levorphenol and epinephrine on stress.
H_a: There is a significant interaction between levorphenol and epinephrine on stress.
$F_{1,16} = 6.05$, $p = 0.0267$.
At $\alpha = 0.05$ we reject H_0 and conclude that there is significant interaction between levorphenol and epinephrine on stress.

5. **a** Yes. There is an apparent "same-direction" interaction, which is reflected in the fact that the difference in mean waiting times between suburban and rural court locations is larger for State 1 than for State 2.

   b <u>Main effect of State:</u> $\quad F_{1,594} = \dfrac{486.0}{2.235} = 217.45$, with $p < 0.001$.

   <u>Main effect of Court Location:</u> $\quad F_{2,594} = \dfrac{413.17}{2.235} = 184.86$, with $p < 0.001$.

   <u>Interaction effect:</u> $\quad F_{2,594} = \dfrac{24.50}{2.235} = 10.96$, with $p < 0.001$.

   All effects are highly statistically significant.

   c Regression model:
   $Y = \beta_0 + \beta_1 S + \beta_2 C_1 + \beta_3 C_2 + \beta_4 SC_1 + \beta_5 SC_2 + E$, where
   $S = 1$ if State 1, -1 if State 2 $\quad C_1 = -1$ if Urban, 1 if Rural, 0 if Suburban
   $C_2 = -1$ if Urban, 1 if Suburban, 0 if Rural

   d We might consider a model of the form
   $Y = \beta'_0 + \beta'_1 S + \beta'_2 C + \beta'_3 SC + E$,
   where S is as defined in (c) and where C is a variable taking on three values (i.e. 0, 1, and 2) which increase directly with the degree of urbanization. One difficulty here is how to determine the appropriate values for C.

9. **a** Here 'Species' should be considered as a fixed factor and 'Locations' as a random factor, unless only the four locations chosen are of interest.

   b

Source	F (Mixed)	p-value	F (Both fixed)	p-value
Species	9.109 (df: 2, 6)	$0.01 < p < 0.025$	0.455 (df: 2, 48)	$p < 0.001$
Locations	1.082 (df: 3, 6)	$p > 0.25$	0.841 (df: 3, 48)	$p > 0.25$
Interaction	1.038 (df: 6, 48)	$p > 0.25$	1.038 (df: 6, 48)	$p > 0.25$

13. **a**

Source	df	SS	F	p
AIRTEMP	3	0.03	0.32	0.813
DOSE	2	0.06	1.04	0.370
A*D	6	0.05	0.29	0.936
Error	24	0.70		
TOTAL	35			

   b From the ANOVA table, we see that the main effects are not significant, and neither is the interaction term.

c An appropriate multiple regression model is:

$$Y = \mu + \sum_{i=2}^{4} \alpha_i X_i + \sum_{j=2}^{3} \beta_j Z_j + \sum_{i=2}^{4}\sum_{j=2}^{3} X_i Z_j \gamma_{ij} + E \quad \text{Where}$$

$$X_i = \begin{cases} -1 & \text{for AIRTEMP} = 21 \\ 1 & \text{for level i of AIRTEMP, i = 2, 3, 4} \\ 0 & \text{otherwise} \end{cases} \qquad Z_j = \begin{cases} -1 & \text{for DOSE} = 0 \\ 1 & \text{for level j of DOSE, j = 2, 3} \\ 0 & \text{otherwise} \end{cases}$$

$\mu = \mu_{..}$
$\alpha_i = \mu_{i.} - \mu_{..}, i = 2, 3, 4$
$\beta_j = \mu_{.j} - \mu_{..}, j = 2, 3$
$\gamma_{ij} = \mu_{ij} - \mu_i + \mu_{..}, i = 2, 3, 4; \quad j = 2, 3$

AIRTEMP = 21 and DOSE = 0 correspond to <u>control</u> levels and hence are of least interest. The larger AIRTEMPS and DOSES are of primary interest so they are directly parameterized in the model.

d A natural polynomial model is
$$Y = \beta_0 + \beta_1 AIRTEMP + \beta_2 DOSE + \beta_3 (AIRTEMP*DOSE) + E.$$
Note that higher order terms can be added to the model if deemed reasonable.

17. a The ANOVA model is:
$Y_{ijk} = \mu + \alpha_i + \beta_j + \gamma_{ij} + E_{ijk}$, $\quad$ i=1,2; $\quad$ j=1,2; $\quad$ k=1,...,6
α_i = effect for the ith school type
β_j = effect for the jth reputation category
γ_{ij} = interaction effect for the ith school type and jth reputation category
E_{ijk} = error term for the observation ijk
The factors are fixed.

b df = 3 $\quad F_{3,20} = 2.572$, with $0.05 < p < 0.10$.

c <u>Main effect of school type:</u>
H_0: There is no significant main effect of school type on starting salary.
H_a: There is a significant main effect of school type on starting salary.

$F_{1,20} = 0.52$, with p = 0.4786.

At $\alpha = 0.05$, we do not reject H_0 and conclude that school type does not have a significant main effect on starting salary.

<u>Main effect of reputation rank:</u>
H_0: There is no significant main effect of reputation rank on starting salary.
H_a: There is a significant main effect of reputation rank on starting salary.

$F_{1,20} = 6.21$, with p = 0.0216.

At $\alpha = 0.05$, we reject H_0 and conclude that reputation rank has a significant main effect on starting salary.

Test of interaction:
- H_0: There is no significant interaction between school type and reputation rank with respect to starting salary.
- H_a: There is significant interaction between school type and reputation rank with respect to starting salary.

$F_{1,20} = 0.99$, with p = 0.3323.

At $\alpha = 0.05$, we do not reject H_0 and conclude that there is not significant interaction between school type and reputation rank.

Chapter 20

1. a Table of sample means:

		Traditional Rank			
		HI	MED	LO	Total
Modern Rank	HI	135	147.5	165	152.5
	MED	145	155	163	155.91
	LO	161.67	143.33	121.67	142.22
	Total	147.22	148.75	154.23	

The above table indicates than the sample mean blood pressure for males with low modern rank is <u>lower</u> than for other modern rank categories; it also illustrates that the sample mean blood pressure for males with low traditional rank is <u>higher</u> than the other traditional rank categories. Finally, the table hints at an interaction effect: for persons with high modern rank, mean blood pressure <u>increases</u> with decreasing traditional rank, whereas for persons with low modern rank mean blood pressure <u>decreases</u> with decreasing traditional rank. In other words, this interaction suggests that persons with incongruous cultural roles (i.e., HI-LO, LO-HI) tend to have higher blood pressures than persons with congruent cultural roles.

b Regression model:

$$Y = \beta_0 + \alpha_1 X_1 + \alpha_2 X_2 + \beta_1 Z_1 + \beta_2 Z_2 + \gamma_{11} X_1 Z_1 + \gamma_{12} X_1 Z_2 + \gamma_{21} X_2 Z_1 + \gamma_{22} X_2 Z_2 + E$$

where

$$X_i = \begin{cases} -1 & \text{if LO modern rank} \\ 1 & \text{if modern rank } i \quad (i = 1 \text{ for HI, } i = 2 \text{ for MED}) \\ 0 & \text{otherwise} \end{cases}$$

and

$$Z_i = \begin{cases} -1 & \text{if LO traditional rank} \\ 1 & \text{if traditional rank } i \ (i = 1 \text{ for HI, } i = 2 \text{ for MED}) \\ 0 & \text{otherwise} \end{cases}$$

c Modern main effect:

(i) $F(X_1, X_2)_{2,27} = \dfrac{\frac{977.70}{2}}{\frac{6358.8746}{27}} = 2.076$, with $0.10 < p < 0.25$ which is not significant.

(ii) $F(X_1, X_2 | Z_1, Z_2)_{2,25} = \dfrac{\frac{872.92}{2}}{\frac{6163.3522}{25}} = 1.768$, with $0.10 < p < 0.25$ which is not significant.

Traditional main effect:

(i) $F(Z_1, Z_2)_{2,27} = \dfrac{\frac{301.30}{2}}{\frac{7035.2732}{27}} = 0.578$, with $p > 0.25$ which is not significant.

(ii) $F(Z_1, Z_2 | X_1, X_2)_{2,25} = \dfrac{\frac{195.52}{2}}{\frac{6163.3522}{25}} = 0.397$, with p > 0.25 which is not significant.

Interaction:

$F(X_1Z_1, X_1Z_2, X_2Z_1, X_2Z_2 | X_1, X_2, Z_1, Z_2)_{4,21} = \dfrac{\frac{4570.85}{4}}{75.8333} = 15.069$ with p < 0.001 which is highly significant.

d $Y = \beta_0 + \beta_1 X_1 + \beta_2 X_2 + \beta_3 X_1 X_2 + E$

where

$X_1 = \begin{array}{ll} 0 & \text{if LO modern rank} \\ 1 & \text{if MED modern rank} \\ 2 & \text{if HI modern rank} \end{array}$ $X_2 = \begin{array}{ll} 0 & \text{if LO traditional rank} \\ 1 & \text{if MED traditional rank} \\ 2 & \text{if HI traditional rank} \end{array}$

The difficulty arises with regard to assigning numerical values to the categories of each factor. The coding scheme for X_1 and X_2 given here assumes that the categories are "equally spaced", which may not really be the case.

5. a Table of means:

		Social Class			
		Lo	Med	Hi	Row means
Number	0	14.571	12.00	13.50	13.5
of times	1	9.00	12.25	7.40	9.4
victimized	2+	8.00	2.33	5.75	5.8
Col means		11.3	9.7	8.8	

The above table suggests a downward trend in confidence with increasing number of victimizations and a slight downward trend with increasing social class status score.

b From the ANOVA table given in the question, we obtain

$F(Victim)_{2,31} = \dfrac{200.00}{22.712} = 8.81$, with p < 0.001 which is significant.

$F(SCLS)_{2,31} = \dfrac{11.37}{22.712} = 0.501$, with p > 0.25 which is not significant.

$F(Interaction)_{4,31} = \dfrac{27.483}{22.712} = 1.21$, with p > 0.25 which is not significant.

c We can use the following regression model:

$$Y = \beta_0 + \alpha_1 X_1 + \alpha_2 X_2 + \beta_1 Z_1 + \beta_2 Z_2 + \gamma_{11} X_1 Z_1 + \gamma_{12} X_1 Z_2 + \gamma_{21} X_2 Z_1 + \gamma_{22} X_2 Z_2 + E$$

where

$$X_1 = \begin{cases} -1 & \text{if no. of times victimized} = 0 \\ 1 & \text{if no. of times victimized} = 1 \\ 0 & \text{otherwise} \end{cases}$$

$$X_2 = \begin{cases} -1 & \text{if no. of times victimized} = 0 \\ 1 & \text{if no. of times victimized} = 2+ \\ 0 & \text{if no. of times victimized} = 1 \end{cases}$$

$$Z_j = \begin{cases} -1 & \text{if social class status} = \text{LO} \\ 1 & \text{if social class status} = j \text{ (j =1 for MED, j = 2 for HI)} \\ 0 & \text{otherwise} \end{cases}$$

We then obtain the following F values from the regression results given in the question:
Main effect of VICTIM:

$$F(X_1, X_2)_{2,37} = \frac{\frac{408.39}{2}}{\frac{836.5834}{37}} = 9.03 \text{, with p < 0.001 which is significant.}$$

$$F(X_1, X_2 | Z_1, Z_2)_{2,35} = \frac{\frac{395.83}{2}}{22.9738} = 8.61 \text{, with p < 0.001 which is significant.}$$

Main Effect of SCLS:

$$F(Z_1, Z_2)_{2,37} = \frac{\frac{45.0673}{2}}{32.4299} = 0.695 \text{, with p > 0.25 which is not significant.}$$

$$F(Z_1, Z_2 | X_1, X_2)_{2,35} = \frac{\frac{32.50}{2}}{22.9738} = 0.707 \text{, with p > 0.25 which is not significant.}$$

Interaction:

$$F(X_1 Z_1, X_1 Z_2, X_2 Z_1, X_2 Z_2 | X_1, X_2, Z_1, Z_2)_{4,31} = \frac{\frac{100.00}{4}}{22.7123} = 1.10 \text{, with p > 0.25 which}$$

is not significant.

d The error term may have a non-constant variance.

9. a A tabulation of DOSAGE * SEX sample sizes reveals that all combinations of these factors have 10 subjects except for 9 in the DOSAGE = 0, SEX = Male combination. We could call this a nearly orthogonal design since we would only need one more observation in the cell with 9 to make a perfectly orthogonal design (equal cell sizes).

    b

Source	TYPE I F	p-value	TYPE II F	p-value
DOSAGE	2.65	$0.05 < p < 0.10$	2.65	$0.05 < p < 0.10$
SEX	0.22	$p > 0.25$	0.22	$p > 0.25$
DOSAGE*SEX	0.14	$p > 0.25$	0.14	$p > 0.25$

    c   At $\alpha = 0.05$, no effects are significant.
    d   Since no effects are significant, no multiple-comparison tests are justified.

13. **a** $Y_{ijk} = \mu + \alpha_i + \beta_j + \gamma_{ij} + E_{ijk}$, $i=1,2,3$; $j=1,2,3$; $k=1,\ldots,n_{ij}$;
        $n_{11} = 3$, $n_{12} = 6$, $n_{13} = 11$, $n_{21} = 4$, $n_{22} = 5$, $n_{23} = 6$, $n_{31} = 4$, $n_{32} = 4$, $n_{33} = 8$
        α_i = effect for the ith clear zone
        β_j = effect for the jth baseline curvature category
        γ_{ij} = interaction effect for the ith clear zone and jth curvature
        E_{ijk} = error term for the observation ijk

    b   df = 8 $F_{8,42} = 3.595$ $0.001 < p < 0.005$

    c   Using the SAS output:
        At $\alpha = 0.10$ we would conclude that there is a significant interaction between CLRZONE and BASECURV. Following the strategy of figure 20-2 in the text, we conclude that the interaction and both main effects are important. (Note that, using the Type III SS provided, we could also conclude that each of the main effects of CLRZONE and BASECURV are significant, given that the other effect *and the interaction* are in the model).

Chapter 21

1. a Days is a crossover factor, since every subject is measured at every level of this factor.
    b The ANOVA model for analyzing the data is $Y_{ij} = \mu + S_i + \tau_j + E_{ij}$, where $i = 1,..., 19$ and $j = 1, 2, 3$, μ is the overall mean, S_i is the random effect of subject i, τ_j is the fixed effect of day j, and E_{ij} is the random error for Day j within subject i. This assumes $\{S_i\}$ and $\{E_{ij}\}$ are mutually independent and that S_i is distributed as $N(0, \sigma_S^2)$ and E_{ij} is distributed as $N(0, \sigma^2)$.
    c $H_0: \tau_1 = \tau_2 = \tau_3 = 0$
    d There seems to be a larger difference between Friday (15.78), and the other days (18.36 and 18.25).
    e The F statistic is given by the expression $F_{t-1, (t-1)(s-1)} = \dfrac{MS_T}{MS_{TS}}$, and this statistic has 2 and 36 d.f. under H_0. The computed value of this F is $40.4973 / 25.8839 = 1.56$, which has a p-value equal to .22, which indicates non-significance.
    f $E[MS_T] = \sigma^2 + \dfrac{19}{2} \sum_{j=1}^{3} \tau_j^2$. Under H_0, this reduces to σ^2.
    g $\hat{\rho} = (367.5946 - 25.8839) / (367.5946 + 25.8839) = .86$

5. a Factor A is a nest factor since each rat is observed at only one level of Factor A.
    b The sample means range from .76 to 2.34, which indicates a fairly large difference between different levels of Factor A, leading to the suggestion that there is an effect of Factor A.
    c The ANOVA model is $Y_{ijk} = \mu + S_{i(j)} + \tau_j + E_{k(ij)}$, where $i = 1,..., 24$; $j = 1,..., 6$; and $k = 1, 2, 3, 4$. μ is the overall mean, τ_j is the fixed effect of Factor A level j, $S_{i(j)}$ is the random effect of rat i within level j of Factor A, and $E_{k(ij)}$ is the random error for repeat k on rat i within level j of Factor A. This model assumes $\{S_{i(j)}\}$ and $\{E_{k(ij)}\}$ are mutually independent and that $S_{i(j)}$ is distributed as $N(0, \sigma_S^2)$ and $E_{k(ij)}$ is distributed as $N(0, \sigma^2)$.
    d Test for Factor A: $F_{5, 18} = \dfrac{MS_A}{MS_{S(A)}} = \dfrac{6.56}{0.046} = 143.0641$, which has a p-value of 0.0001 (highly significant).
    e $E[MS_A] = \sigma^2 + 4\sigma_S^2 + \dfrac{(24)(4)}{5} \sum_{j=1}^{6} \tau_j^2$. Under H_0: all $\tau_j = 0$, $E[MS_A]$ reduces to $\sigma^2 + 4\sigma_S^2$.

9. a The ANOVA model: $Y_{ijk} = \mu + S_i + \alpha_j + \beta_k + \delta_{jk} + S_{ij} + S_{ik} + E_{ijk}$. This model assumes $\{S_i\}$, $\{S_{ij}\}$, $\{S_{ik}\}$, and $\{E_{ijk}\}$ are mutually independent, and that S_i is distributed as $N(0, \sigma_S^2)$, S_{ij} is distributed as $N(0, \sigma_{SD}^2)$, S_{ik} is distributed as $N(0, \sigma_{ST}^2)$, and E_{ijk} is distributed as $N(0, \sigma^2)$, where D denotes the Days factor and T denotes the Time factor.
    b Neither Days (F = 1.8226, P = 0.1762), nor Time (F = 2.2986, P = 0.1469), nor their interaction (F = 0.3324, P = 0.7194) are significant.

c Expected mean squares:

(Both Factors Fixed)

SOURCE	MS	E(MS)
Days	MS_D	$\sigma^2 + 2\sigma^2_{SD} + \frac{38}{2}\sum_{j=1}^{3} \alpha_j^2$
Time	MS_T	$\sigma^2 + 3\sigma^2_{ST} + \frac{57}{2}\sum_{k=1}^{2} \beta_k^2$
Days*Time	MS_{DT}	$\sigma^2 + \frac{19}{2}\sum_{j=1}^{3}\sum_{k=1}^{2} \delta_{jk}^2$

SOURCE	H_0	What each E(MS) reduces to under H_0
Days	$\alpha_j = 0$	$\sigma^2 + 2\sigma^2_{SD}$, which is the E(MS) for MS_{SD}
Time	$\beta_k = 0$	$\sigma^2 + 3\sigma^2_{ST}$, which is the E(MS) for MS_{ST}
Days*Time	$\delta_{jk} = 0$	σ^2, which is the E(MS) for MS_{Error}

d The correlation structure for this model is not exchangeable. The correlation between responses on different days but at the same time of day is different from the correlation between responses on different days at different times, and each of these is different from the correlation between responses on the same day but at different times.

Chapter 22

1. a Yes they are identical if one is able to assume that the linear regression model is fit to normally distributed data.

b For the model $Y = \beta_0 + \beta_1 DRUG + \beta_2 SBP + \beta_3 QUET + E$
where DRUG = 1 if drug, 0 if placebo
$H_0: \beta_1 = 0 \quad H_a: \beta_1 \neq 0$

Test Statistic: $Z = \dfrac{\hat{\beta}_1}{\sqrt{Var_{\hat{\beta}_1}}}$ with approximately a standard normal distribution.

Critical Value: if $Z > 1.96$ then reject H_0 at $\alpha = 0.05$.

c No, however as N increases they approach one another.

d $H_0: \beta_1 = 0 \quad H_a: \beta_1 \neq 0$
Test Statistic: $-2 \log L \text{ (reduced)} - (-2 \log L \text{ (full)}) = \chi^2$ with 1 df where

full model equals: $Y = \beta_0 + \beta_1 DRUG + \beta_2 SBP + \beta_3 QUET + E$ and

reduced model equals: $Y = \beta_0 + \beta_2 SBP + \beta_3 QUET + E$

The test statistic approximates a chi-square distribution with 1 df.
Critical Value: if $\chi^2 > 3.841$ then reject H_0 at $\alpha = 0.05$.

e Yes since $\chi^2 = Z^2$ when Z is normally distributed.

f 95% confidence interval for β_1: $\hat{\beta}_1 \pm 1.96 * \sqrt{Var_{\hat{\beta}_1}}$

Chapter 23

1.

a OR = 2.925 so to estimate $\hat{\beta}_1$ we take ln 2.925 which equals 1.073.

b logit [pr (Y = 1)] = $-2.8 + 0.706X_1 + 0.0004X_2 + 0.0006X_3$

c $[pr(Y=1)] = \dfrac{1}{1 + e^{-(-2.8 + 0.706(1) + 0.0004(20) + 0.0006(20)(1))}} = \dfrac{1}{1 + e^{-(-2.074)}} = 0.112$

d $\hat{OR}_{20 \text{ yr old smoker vs } 21 \text{ yr old smoker}} = e^{-(\hat{\beta}_2 + \hat{\beta}_3)} = e^{-(0.0004 + 0.0006)} = 0.999$

A 21 year-old smoker has an odds of 1.001 times greater for hypertension relative to a 20 year-old smoker. The association is essentially the same for the two ages.

e 95% CI for $OR_{20 \text{ yr old smoker vs } 21 \text{ yr old smoker}}$: $-(\hat{\beta}_2 + \hat{\beta}_3) \pm 1.96\sqrt{(-1)^2 \hat{Var}(\hat{\beta}_2) + (-1)^2 \hat{Var}(\hat{\beta}_3) + 2(-1)(-1)\hat{cov}(\hat{\beta}_2, \hat{\beta}_3)}$

$e^{-0.0004 - 0.0006 \pm 1.96\sqrt{1 \times 10^{-8} + 9 \times 10^{-8} + 2(3 \times 10^{-8})}} = (0.9982, 0.9998)$

f $H_0: \beta_3 = 0 \quad H_a: \beta_3 \neq 0$

Test Statistic: $-2 \log L \text{ (reduced)} - (-2 \log L \text{ (full)}) = \chi^2$ with 1 df.

$\chi^2 = 308.00 - 303.84 = 4.16$, with $0.025 < p < 0.05$.

At $\alpha = 0.05$ we reject H_0 and conclude that there is significant interaction between age and smoking.

Chapter 24

1.
- **a** $n = 30$ (6 age-sex groups * 5 years)
- **b** Here, $i = 5$ and $k = 1962 - 1960 = 2$, so that
$$E(Y_{52}) = \iota_{52}\lambda_{52} = \iota_{52}\, e^{(\alpha_5 + 2\beta)}$$
- **c** Log rate changes linearly with time. In particular, for the i-th group,
$$\ln \lambda_{ik} = \alpha_i + \beta k$$
so that α_i is the intercept and β is the slope of the straight line relating the response $\ln \lambda_{ik}$ to the time variable $k = $ [year] $- 1960$.
- **d** Model (1) assumes no interaction between age-sex group and time in the sense that the change in log rate over time (as measured by β) does not depend on i. Since $\ln \lambda_{ik} = \alpha_i + \beta k$, a graph of $\ln \lambda_{ik}$ versus k for each i would plot as a series of parallel straight lines, that is, lines all with the same slope (β) but possibly different intercepts (the α_i's). A lack of parallelism would reflect interaction between age-sex groups and time because the change in log rate over time would be different for different age-sex groups.
- **e** $lnIDR_{ik} = \ln\lambda_{ik} - \ln\lambda_{10} = (\alpha_i + \beta k) - (\alpha_1 + \beta*0) = (\alpha_i - \alpha_1) + \beta k$, so that
$$IDR_{ik} = e^{\alpha_i - \alpha_1} e^{\beta k}$$
Note that this is a function of both age-sex group (i) and time (k).
- **f** An appropriate model is
$$E(Y_{ik}) = \iota_{ik}\lambda_{ik}$$
where
$$\ln\lambda_{ik} = \sum_{i=1}^{6} \alpha_i A_i + \beta k + \sum_{i=1}^{5} \gamma_i(A_i k) \ .$$

For age-sex group i, then,
$$\ln \lambda_{ik} = \alpha_i + \beta k + \gamma_i k$$
$$= \alpha_i + (\beta + \gamma_i)k$$
$$= \alpha_i + \delta_i k$$
where $\delta_i = \beta + \gamma_i$. Hence the slope for group i, namely δ_i, is now a function of i. Only when all six δ_i's (or, equivalently, all six γ_i's) are equal will the straight lines be parallel.
- **g** Yes, since $D(\hat{\beta})_{(1)} - D(\hat{\beta})_{(3)} = 300 - 175 = 125$, which is highly significant when compared to appropriate upper tail χ^2-values with $29 - 24 = 5$ df.
- **h** Yes, since $D(\hat{\beta})_{(3)} - D(\hat{\beta})_{(4)} = 175 - 60 = 115$, which is highly significant when compared to appropriate upper tail χ^2-values with $24 - 23 = 1$ df.
- **i** No, since $D(\hat{\beta})_{(4)} - D(\hat{\beta})_{(5)} = 60 - 59 = 1$, which is clearly not significant when compared to appropriate upper tail χ^2-values with $23 - 22 = 1$ df.
- **j** Yes, since $D(\hat{\beta})_{(4)} - D(\hat{\beta})_{(7)} = 60 - 20 = 40$, which is highly significant when compared to appropriate upper tail χ^2-values with $23 - 18 = 5$ df.
- **k** H_0 is rejected, since $D(\hat{\beta})_{(4)} - D(\hat{\beta})_{(6)} = 60 - 25 = 35$, which is highly significant when compared to appropriate upper tail χ^2-values with $23 - 22 = 1$ df.